International Conference on Materials Engineering and Nanotechnology (ICMEN 2019)

IOP Conference Series: Materials Science and Engineering
Volume 839

Kuala Lumpur, Malaysia
2 - 5 December 2019

ISBN: 978-1-7138-2681-1
ISSN: 1757-8981

Printed from e-media with permission by:

Curran Associates, Inc.
57 Morehouse Lane
Red Hook, NY 12571

Some format issues inherent in the e-media version may also appear in this print version.

Institute of Physics
Dirac House, Temple Back
Bristol BS1 6BE UK

Phone: 44 1 17 929 7481
Fax: 44 1 17 920 0979

techtracking@iop.org

Additional copies of this publication are available from:

Curran Associates, Inc.
57 Morehouse Lane
Red Hook, NY 12571 USA
Phone: 845-758-0400
Fax: 845-758-2633
Email: curran@proceedings.com
Web: www.proceedings.com

TABLE OF CONTENTS

IOP Conf. Series: Materials Science and Engineering **839** (2020) 011001 doi:10.1088/1757-899X/839/1/011001

Preface

The International Conference on Materials Engineering and Nanotechnology (ICMEN 2019) was held in Kuala Lumpur, Malaysia during Decemebr 2-5, 2019. The aim as well as objective of ICMEN 2009 is to present the latest research and results of scientists related to Materials Engineering and Nanotechnology topics. The conference provides a useful and wide platform both for display the latest research and for exchange of research results and thoughts in Advanced Materials Research and Nanotechnology. The participants of the conference were from almost every part of the world, with background of either academia or industry, even well-known enterprise.

As a platform for technical exchange, each author has 20 minutes to introduce his or her research. In addition to the presentation, each author is encouraged to connect and exchange ideas with other scholars. The success and prosperity of the conference is reflected high level of the papers received. Each contributed paper has been peer-reviewed by reviewers who were collected organizing and technical committee members as well as other experts in the field from different countries.

We would like to acknowledge all of those who supported ICMEN 2019. Each individual and institutional help were very important for the success of this conference. Especially we would like to thank the organizing committee for their valuable advices in the organization and helpful peer review of the papers.

We sincerely hope that ICMEN 2019 will be a forum for excellent discussions that will put forward new ideas and promote collaborative researches. We are sure that the proceedings will serve as an important research source of references and the knowledge, which will lead to not only scientific and engineering progress but also other new products and processes.

IOP Conf. Series: Materials Science and Engineering **839** (2020) 011002 doi:10.1088/1757-899X/839/1/011002

Peer review statement

All papers published in this volume of *IOP Conference Series: Materials Science and Engineering* have been peer reviewed through processes administered by the proceedings Editors. Reviews were conducted by expert referees to the professional and scientific standards expected of a proceedings journal published by IOP Publishing.

IOP Conf. Series: Materials Science and Engineering **839** (2020) 012001 doi:10.1088/1757-899X/839/1/012001

Study on the effect of ZIF-8 nanoparticles on the transport behavior of CH_4 and CO_2 gases in polysulfone membrane

Christine Ch'ng Wei Mann[1,2], *Yin Fong Yeong[1,2] and Norwahyu Jusoh[1,3]

[1] Chemical Engineering Department, Universiti Teknologi PETRONAS, 32610 Seri Iskandar, Perak, Malaysia.
[2] CO_2 Research Centre (CO_2RES), Institute of Contaminant Management, Universiti Teknologi PETRONAS, 32610 Seri Iskandar, Perak, Malaysia.
[3] Centre for Contaminant Control & Utilization (CenCoU), Institute of Contaminant Management, Universiti Teknologi PETRONAS, 32610 Seri Iskandar, Perak, Malaysia.

*Corresponding author: yinfong.yeong@utp.edu.my

Abstract. CO_2 capture has been given much attention worldwide in order to minimize the CO_2 emission in protecting the harmful impact on climate change. In this study, zeolitic imidazolate framework-8 (ZIF-8) nanoparticles was embedded in polysulfone (PSf) polymer and the changes of the transport behavior of CO_2 and CH_4 single gases of the resultant membranes were investigated. The solubility of the gases through the membranes were determined by using dual-volume sorption cell at 5 bar and room temperature. Subsequently, the diffusion coefficients were determined from the permeability and solubility values. From the results obtained, the CO_2 sorption values for both resultant membranes were higher than that of CH_4. Meanwhile, the membrane incorporated with ZIF-8 demonstrated greater CO_2 sorption capacity compared to pure PSf membrane. It can be concluded that the improvement of CO_2 permeability after the incorporation of ZIF-8 into PSf is mainly due to the enhancement in solubility and diffusivity of the membrane.

1. Introduction

In recent years, membrane-based technology has progressively dominating the gas separation applications as an alternative to the existing extensive technologies such as cryogenic distillation and pressure swing adsorption. Membrane process is preferable due to its distinctive properties of space and weight saving, requires minimum maintenance and costing as well as easy installation onto existing equipment [1, 2]. Polymeric membrane is known to demonstrate trade-off performance between gas permeability and gas pair selectivity [3, 4]. Thus, the addition of nanomaterials in forming hybrid membranes is an advancement to increase the gas permeability and gas selectivity performance.

ZIFs is a nano-porous material with distinct pore structures with desirable properties in terms of high porosity and exceptional thermal and chemical stability [5, 6]. ZIF-8 is chosen in this study as it possesses high selective adsorption ability towards CO_2 [7]. Thus, the incorporation of ZIF-8 particles into PSF polymer will increase the affinity towards CO_2 while increasing the free volume of the polymer structure due to the alteration in the polymer chain packing [8].

It is of interest to focus on varying polymer structure to increase permeability while maintaining its selectivity. In this work, correlation between the polymer structure with its free volume and ZIF-8 particles have been made. It is expected for the ZIF-8 hybrid membrane to reduce the chain mobility which leads to higher selectivity. Besides, the change of molecular structure in the polymer can improve the trade-off performance to a certain extent.

This present work involves the synthesis of pure PSf and ZIF-8 incorporated hybrid membrane. The transport properties including permeability, solubility and diffusivity of CO_2 and CH_4 gases via gas permeation and sorption tests for the hybrid membrane has been investigated and compared with the values obtained for pure PSf. This information is required to gain insights into the effect of incorporating ZIF-8 particles towards changes in gas transport properties of CO_2 and CH_4 through the PSf polymer. These results will gain a better fundamental understanding of this hybrid nanoparticles and polymer for the ease of their potential implementation in industrial gas separation.

2. Methodology

2.1. Materials

Zinc nitrate hexahydrate ($Zn(NO_3)_2.6H_2O$, > 98%, Aldrich), 2-methylimidazole (2-MeIM, > 99%, Aldrich), polysulfone pellets (Mn ~ 35000, Merck), methanol (CH_3OH, > 99.9%, Merck) and chloroform ($CHCl_3$ > 99.8%, Merck) were used as received. Carbon dioxide (CO_2, 99.95%) and methane (CH_4, 99.95%) gases were supplied by Air Product Malaysia Sdn. Bhd.

2.2. Preparation of ZIF-8

Two solutions containing 2-methylimidazole in 200 mL of methanol and zinc nitrate hexahydrate in 200mL of methanol were prepared, respectively. The two solutions were subsequently mixed rapidly and stirred for 1 h. The solution will eventually turn cloudy and it is then proceeded with centrifugation at 7800 rpm for an interval of 5 min to obtain the ZIF-8 particles from the solution mixture. The ZIF-8 particles were then washed with methanol for 3 times and placed in an oven at 85 °C for 24 h. Lastly, the dried ZIF-8 particles were grinded into finer particles.

2.3. Fabrication of membranes

In this work, pure PSf flat sheet membrane was fabricated using solution casting method as reported in the literature [9]. PSf pellets were firstly dried in an oven at 60 °C to remove excessive moisture. Consecutively, PSf pellets was dissolved in chloroform in a sealed glass bottle and left for 8 h under stirring at room temperature. The polymer solution formulation was set at 10% w/v. The resultant PSf solution was left for degassing overnight at room temperature. The PSf solution was subsequently casted onto a petri dish and covered with a glass plate, then left for 24 h to allow slow solvent evaporation at room temperature. Finally, the membrane film was peeled off from petri dish and dried at 60 °C for 8 h in vacuum oven to remove the remaining solvent.

ZIF-8/PSf hybrid membrane was also fabricated by using solution blending method [9, 10]. 0.981 g of PSF was dissolved in chloroform in a sealed glass bottle. The polymer solution was left stirring for 8 h at room temperature. In the meantime, 0.009 g of ZIF-8 was added into chloroform solvent in a different sealed glass bottle. The filler suspension was alternately stirred and sonicated for 1 h to disperse the particle. Subsequently, priming method was adapted where 10 % of the PSf solution was added into the filler-solvent suspension, followed by alternate stirring and sonication for 1 h. The step was repeated after adding the remaining 90 % of the PSf solution into filler-solvent suspension. The final mixture solutions were vigorously stirred for 1 h before casted onto the petri dish. The mixture solutions were left at room temperature for 24 h to allow solvent evaporation. The hybrid membrane consisting 1 wt% loading of ZIF-8 was labelled as ZIF-8-1.0/PSf.

2.4. Characterization of Membranes

The morphology of the fabricated membranes was observed under Zeiss Supra 55 VP field emission scanning electron microscope (FESEM). The cross-sections membranes were obtained via freeze-fracturing under liquid nitrogen and sputter-coating with Pt using Quorum Q150R S sputter coater. The density of the membrane samples was measured using Mettler Toledo, OHAUS CP224C density kit. The weight of the membrane samples was first measured in air and high purity ethanol, respectively and the results were then used to determine the density of membrane samples. The fractional free volume (FFV) of the membranes was calculated using equations (3-4) as follows [10] :

$$V = \frac{M}{p} \quad (1)$$

$$FFV = \frac{V - 1.3V_w}{V} \quad (2)$$

where V is the molar volume of the repeating unit of polymer (cm^3/mol), M is to the molecular weight of the repeating unit of polymer (g/mol), ρ is the density of the membrane (g/cm^3) and V_w is the van der Walls molar volume (cm^3/mol) via Bondi method.

2.5 Transport Behavior of Gases

2.5.1 Gas Permeation

Single gas permeation tests were carried out using a custom-built gas test rig described in detailed from our previous reported work [11].The CO_2 and CH_4 single permeation tests were performed at constant feed pressure of 5 bar and constant flow rate of 300 ml/min at room temperature. The gas permeability was calculated using equation (3) as follows [12]:

$$P = \frac{QL}{A(\Delta P)} \quad (3)$$

where P is the permeability of gases of membrane sample (Barrer), Q *is the* volumetric flow rate of gas through the membrane sample (cm^3/s), L is the membrane thickness (cm), A is the membrane area (cm^2) and ΔP is the pressure drop across the membrane (cmHg), respectively.

The CO_2/CH_4 ideal selectivity of the membrane was determined by calculating the ratio of CO_2 permeability and permeability of CH_4 as follows:

$$\alpha_p = \frac{P_{CO_2}}{P_{CH_4}} \quad (4)$$

where α_p specifies the selectivity of CO_2/CH_4 and P refers to the permeability (Barrer).

2.5.2 Gas Sorption

The solubilities of CO_2 and CH_4 were measured where pressure decay principle on a dual-chamber setup is applied [13]. The dual-chamber set-up consists of a sample chamber and a reservoir chamber located next to each other interconnected by a valve. The testing was conducted at a pressure of 5 bar at room temperature.

Prior to testing, the mass of the membrane sample was recorded. The membrane sample is then placed into the sample chamber and the whole set-up was vacuumed overnight. Subsequently, the reservoir chamber was filled with gas of 5 bar. The amount of gas, n_1 in the reservoir chamber can be calculated using equation (5), as follows [14] :

$$n_1 = \frac{5V_{C1}}{Z_1 RT} \tag{5}$$

where V_{C1} is the volume of the chamber, T is temperature of the gas, Z_1 is compressibility factor of the real gas under *5 bar* and T conditions and R is the ideal gas constant, respectively.

The valve that interconnects both the sample and reservoir chamber was then opened and the gas starts to flow into the sample chamber and the sorption process by the polymer starts to take place. Once equilibrium point is reached, the final pressure, P_2 remains constant, and the amount of gas remaining in the reservoir and membrane sample chambers, can be calculated using equation (6), as follows [14] :

$$n_{1+2} = \frac{P_2(V_{C1} + V_{C2} - V_S)}{Z_2 RT} \tag{6}$$

where V_{C2} is the volume of the membrane chamber, V_s is volume of the membrane sample, Z_2 is the compressibility factor at P_2 and T.

The amount of gas dissolved into the polymer, n_p, can be calculated using equation (7) as follows [14] :

$$n_p = n_1 - n_{1+2} \tag{7}$$

The concentration of the gas molecule adsorbed into the membrane was calculated using equation (7) as follows [13]:

.

$$C = \frac{n}{Vp} \text{ x } 22414 \tag{8}$$

where 22414 cm^3/mol is a numerical conversion factor and V_p is volume of the polymer sample in the membrane chamber (cm^3)

The solubility of the membrane was calculated using equation (8) as follows [15]:

$$S = \frac{C}{P} \tag{9}$$

where S is the solubility (cm^3 (STP)/cm^3.bar) and P is the pressure of the gas (bar), respectively.

The solubility selectivity was calculated using equation (9) as follows:

$$\alpha_{S,CO_2/CH_4} = \frac{S_{CO_2}}{S_{CH_4}} \tag{10}$$

where α_s specifies the solubility se lectivity of CO_2/CH_4.

2.5.3 Gas Diffusivity

The diffusivity of the membrane was calculated using equation (10) as follows [15] :

IOP Conf. Series: Materials Science and Engineering **839** (2020) 012001 doi:10.1088/1757-899X/839/1/012001

$$D = \frac{P}{S} \tag{11}$$

where *D*, *P* and *S* are the diffusivity (cm^2/s), permeability (cm^3 (STP).cm/s.cm^2.cmHg) and solubility (cm^3 (STP)/ cm^3.cmHg), respectively.

The diffusivity selectivity was calculated using equation (11) as follows:

$$\alpha_{D,CO_2/CH_4} = \frac{D_{CO_2}}{D_{CH_4}} \tag{12}$$

where α_D specifies the diffusivity selectivity of CO_2/CH_4.

3. Results and Discussion

3.1. Characterization of Membranes

Figure 1 demonstrates the morphology of membranes at cross-sectional view of FESEM images. It can be observed that the ZIF-8 particles were well contained in the PSf polymer. There was no significant formation of interfacial voids between the particles and polymer phase. Overall, the incorporation of ZIF-8 particles has exhibited a good dispersion of particles in polymer matrix and is expected to improve the performance of gas permeation in the hybrid membrane.

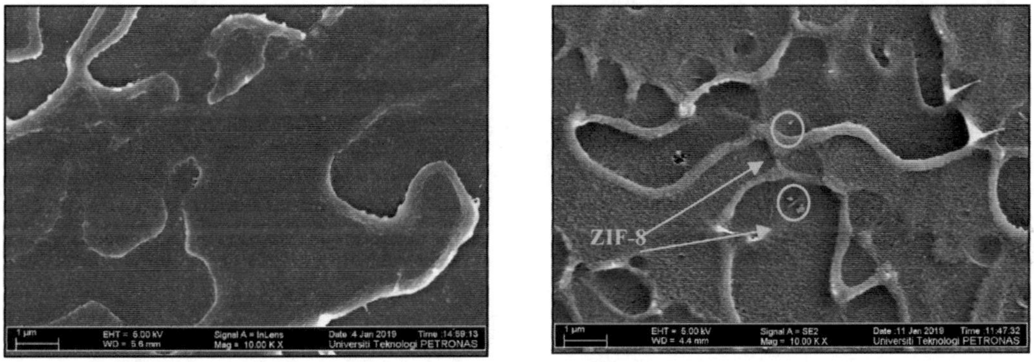

Figure 1. Cross section view FESEM images of membranes (a) pure PSf (b) ZIF-8-1.0/PSf

3.2 FFV of Membranes

Table 1 demonstrates the densities and FFVs of the resultant membranes. The density and FFV of 1.26 g/cm^3 and 0.14 were obtained for pure PSf membrane whereas 1.28 g/cm^3 and 0.16 were obtained for ZIF-8-1.0/PSf membrane. From Table 1, the hybrid membrane demonstrated lower density and higher FFV in comparison to the pure PSf membrane. The results shown demonstrated that the incorporation of ZIF-8 has led to the increment of accessible cavities in the PSf phase, which increases the FFV value.

Table 1. Density and FFV of fabricated pure PSf and ZIF-8/PSf membranes

Membranes	Density (g/cm^3)	Fractional free volume (FFV)
Pure PSf	1.28	0.14
ZIF-8-1.0/PSf	1.26	0.16

3.3 Gas Permeation Performance

Permeability measurements were carried out for CO_2 and CH_4 gases at room temperature and 5 bar. Pure gas permeation and selectivity results for CO_2 and CH_4 were shown in Table 2 for pure PSf and ZIF-8-1.0/PSf membranes. CO_2 and CH_4 permeabilities of 6.28 Barrer and 0.27 Barrer were obtained for pure PSf membrane. Meanwhile, CO_2 permeability at 33.12 Barrer was obtained for ZIF-8-1.0/PSf hybrid membrane with CO_2/CH_4 selectivity of 38.8. From the results obtained in Table 2, it was found that ZIF-8-1.0/PSf exhibited higher permeability and selectivity compared to pure PSf membrane. In comparison to the pure PSF membrane, an increment of 527% in CO_2 permeability with the highest CO_2/CH_4 ideal selectivity of 38.8 is achieved for ZIF-8-1.0/PSf. Overall, the addition of ZIF-8 into PSf polymer has resulted in the increment of separation performance of the membrane. The increment of permeability is in agreement with the enhancement of FFV values in the hybrid membrane. The significant improvement of CO_2 permeability is related to the transportation of gas molecules through solution-diffusion mechanism/filler pores through the polymer matrix [16]. Therefore, non-selective voids in the hybrid membrane could be reduced and thus, the gas selectivity increases. The membranes are then proceeded with sorption and diffusivity study.

Table 2. Permeabilities of pure PSf and ZIF-8/PSf membranes

Membranes	Permeability, P (Barrer)		Selectivity, α
	CO_2	CH_4	P_{CO2}/P_{CH4}
Pure PSf	6.28	0.27	23.3
ZIF-8-1.0/PSf	33.12	0.85	38.8

3.4 Gas Solubility

Gas solublity of CO_2 and CH_4 for the resultants membranes at pressure of 5 bar at room temperature is shown in Table 3. The CO_2 and CH_4 solubilities of $35.23x10^{-2}$ $cm^3(STP)/cm^3.cmHg$ and $10.51x10^{-2}$ cm^3 $(STP)/cm^3.cmHg$ were obtained for pure PSf while $71.43x10^{-2}$ $cm^3(STP)/cm^3.cmHg$ to 16.25 $x10^{-2}$ $cm^3(STP)/cm^3.cmHg$ were obtained for ZIF-8-1.0/PSf, respectively. CO_2/CH_4 solubility selectivity ranging from 3.35 to 4.40 was also attained for the resultant membranes, where ZIF-8-1.0/PSf membrane demonstrates the highest CO_2/CH_4 solubility selectivity.

Table 3. Solubilities of fabricated pure PSf and ZIF-8/PSf membranes

Membranes	Solubility, S (10^{-2} $cm^3(STP)/cm^3.cmHg$)		Selectivity, α
	CO_2	CH_4	S_{CO2}/S_{CH4}
Pure PSf	35.23	10.51	3.35
ZIF-8-1.0/PSf	71.43	16.25	4.40

From the results obtained, it is shown the inclusion of ZIF-8 particles into PSf membrane has greater CO_2 sorption capability compared to pure PSf membrane. This is correlated with the increase of free volume of polymer structure and the nature of ZIF-8 particles which are more condensable with CO_2 gas and thus increases the gas solubility in the membrane. Besides, it can be observed that the CO_2 sorption for ZIF-8 based membranes is higher as compared to the CH_4 sorption, which is in line with the higher condensabilities of CO_2 compared to CH_4. Furthermore, higher critical temperature of CO_2 and attraction of CO_2 molecules towards ZIF-8 particles than that of CH_4 also contributes to the enhancement of CO_2 solubility in the hybrid membrane [17, 18].

3.5 Gas Diffusivity

Table 4 shows the diffusivity, D of CO_2 and CH_4 in pure PSf and ZIF-8-1.0/PSf membrane at pressure and temperature of 5 bar and room temperature.

Table 4. Diffusivities of fabricated pure PSf and ZIF-8/PSf membranes

Membranes	Diffusivity, D (10^{-8} cm^2/s)		Selectivity, α
	CO_2	CH_4	D_{CO2}/D_{CH4}
Pure PSf	1.78	0.26	6.99
ZIF-8-1.0/PSf	3.98	0.53	8.10

Based on Table 4, CO_2 and CH_4 diffusivities of $1.78x10^{-8}$ cm^2/s and 0.26 $x10^{-8}$ cm^2/s were obtained for pure PSf membrane while $3.98x10^{-8}$ cm^2/s and to 0.53^{-8} cm^2/s were obtained for ZIF-8-1.0/PSf, respectively. CO_2/CH_4 diffusivity selectivity ranging from 6.99 to 8.10 are also obtained for the resultant membranes, where ZIF-8-1.0/PSf membranes demonstrates the highest CO_2/CH_4 diffusivity selectivity.

From the results obtained, it is shown that the diffusivity of CH_4 is lower in comparison to CO_2, which is consistent with the known kinetic diameter of CH_4 (3.8 Å) > CO_2 (3.4Å). The embedded ZIF-8 particles in PSf polymer has indirectly created a diffusive path for the CO_2 gas molecule to pass through the membrane with ease, subsequently increased the overall diffusivity [19]. Furthermore, the solubility, permeability and diffusivity of the hybrid membrane are higher than pure PSf membrane, which is consistent with the increment of FFV obtained by ZIF-8 incorporated membrane as discussed in the previous section.

3.6 Performance of Membranes via Robeson's Upper Bound

In this study, the inclusion of ZIF-8 filler into PSF has resulted in the enhancement of gas separation performance. For further verification, Figure 2 shows the comparison of the separation performance of membranes obtain in this work with reported results for PSF-based MMMs through CO_2/CH_4 Robeson's upper bound. ZIF-8-1.0/PSf hybrid membrane fabricated in this work demonstrated higher CO_2 permeability and CO_2/CH_4 ideal selectivity in comparison to CO_2/CH_4 gas pair separation performance of CNTs-PSF [20] and CU_3BTC_2/PSF [21] membranes.

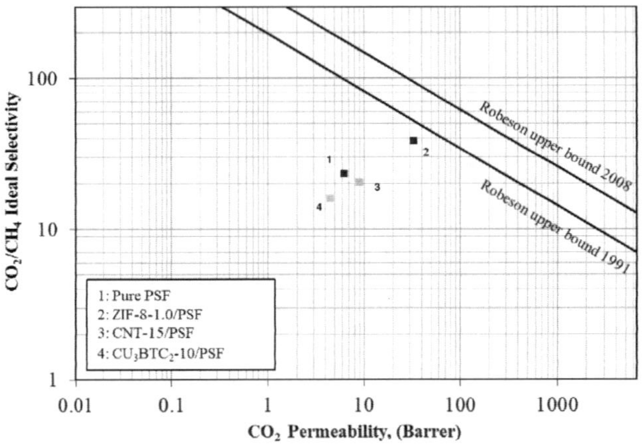

Figure 2. Performance comparison between membranes fabricated in the present work and PSF based hybrid membranes reported in literature via CO_2/CH_4 Robeson's upper bound.

4 Conclusion

Overall, the incorporation of ZIF-8 particles in PSf has led to the improvement of higher permeability, solubility and diffusivity. It is observed that the transportation of gas molecules through the membranes has improved mostly due to increase of free volume and change of polymer structure after ZIF-8 particles was embedded into PSf polymer, which consequently increases the selectively of the hybrid membrane. It can be concluded that the improvement of CO_2 permeability in ZIF-8 based membranes is simultaneous with the enhancement in solubility and diffusivity of the membrane, proving a major role played in both solubility and diffusivity.

Acknowledgement

The authors would like to express acknowledgement for the financial support by Universiti Teknologi PETRONAS and technical support by CO_2 research Centre (CO_2RES), Institute of Contaminant Management.

References

[1] An H, Park S, Kwon H T, Jeong H-K and Lee J S 2017 A new superior competitor for exceptional propylene/propane separations: ZIF-67 containing mixed matrix membranes *Journal of Membrane Science* **526** 367-76
[2] Baker R W 2012 *Membrane technology and applications*: John Wiley & Sons)
[3] Robeson L M 1991 Correlation of separation factor versus permeability for polymeric membranes *Journal of Membrane Science* **62** 165-85
[4] Robeson L M 2008 The upper bound revisited *Journal of Membrane Science* **320** 390-400
[5] Lee M J, Abdul Hamid M R, Lee J, Kim J S, Lee Y M and Jeong H-K 2018 Ultrathin zeolitic-imidazolate framework ZIF-8 membranes on polymeric hollow fibers for propylene/propane separation *Journal of Membrane Science* **559** 28-34

IOP Conf. Series: Materials Science and Engineering **839** (2020) 012001 doi:10.1088/1757-899X/839/1/012001

[6] Chen B, Yang Z, Zhu Y and Xia Y 2014 Zeolitic imidazolate framework materials: recent progress in synthesis and applications *Journal of Materials Chemistry A* **2** 16811-31

[7] Li Y, Wang K, Zhou W, Li Y, Vila R, Huang W, Wang H, Chen G, Wu G-H, Tsao Y, Wang H, Sinclair R, Chiu W and Cui Y 2019 Cryo-EM Structures of Atomic Surfaces and Host-Guest Chemistry in Metal-Organic Frameworks *Matter* **1** 428-38

[8] Sokolova M P, Smirnov M A, Geydt P, Bugrov A N, Ovaska S-S, Lahderanta E and Toikka A M 2016 Structure and Transport Properties of Mixed-Matrix Membranes Based on Polyimides with ZrO_2 Nanostars *Polymers (Basel)* **8** 403

[9] Waheed N, Mushtaq A, Tabassum S, Gilani M, Ilyas A, Ashraf F, Jamal Y, Bilad M, Khan A and Khan A 2016 *Mixed Matrix Membranes Based on Polysulfone and Rice Husk Extracted Silica for CO2 Separation* vol 170

[10] Bondi A 1964 van der Waals Volumes and Radii *The Journal of Physical Chemistry* **68** 441-51

[11] Jusoh N, Yeong Y F, Cheong W L, Lau K K and M. Shariff A 2016 Facile fabrication of mixed matrix membranes containing 6FDA-durene polyimide and ZIF-8 nanofillers for CO2 capture *Journal of Industrial and Engineering Chemistry* **44** 164-73

[12] Mohammadi T, Moghadam M T, Saeidi M and Mahdyarfar M 2008 Acid Gas Permeation Behavior Through Poly(Ester Urethane Urea) Membrane *Industrial & Engineering Chemistry Research* **47** 7361-7

[13] Chakma A and Feng X 2006 Sorption, diffusion, and permeation of light olefins in poly(ether block amide) membranes *Chemical Engineering Science* **61** 6142-53

[14] Koros W J, Paul D R and Rocha A A 1976 Carbon dioxide sorption and transport in polycarbonate **14** 687-702

[15] Lin W-H and Chung T-S 2001 Gas permeability, diffusivity, solubility, and aging characteristics of 6FDA-durene polyimide membranes *Journal of Membrane Science* **186** 183-93

[16] Waqas Anjum M, Bueken B, De Vos D and Vankelecom I F J 2016 MIL-125(Ti) based mixed matrix membranes for CO2 separation from CH4 and N2 *Journal of Membrane Science* **502** 21-8

[17] Wang Y-C, Huang S-H, Hu C-C, Li C-L, Lee k-r, Liaw D-J and Lai J-Y 2005 Sorption and transport properties of gases in aromatic polyimide membranes *Journal of Membrane Science* **248** 15-25

[18] Azizi N, Mohammadi T and Behbahani R 2016 Comparison of permeability performance of PEBAX-1074/TiO2, PEBAX-1074/SiO2 and PEBAX-1074/Al2O3 nanocomposite membranes for CO2/CH4 separation *Chemical Engineering Research and Design* **117** 177-89

[19] Escorihuela Roca S, Valero L, Tena A, Shishatskiy S, Escolástico S, Brinkmann T and Serra J 2018 Study of the Effect of Inorganic Particles on the Gas Transport Properties of Glassy Polyimides for Selective CO2 and H2O Separation *Membranes* **8** 128

[20] Kim S, Chen L, Johnson K and Marand E 2007 Polysulfone and functionalized carbon nanotube mixed matrix membranes for gas separation: Theory and experiment *Journal of Membrane Science* **294** 147-58

[21] Car A, Stropnik C and Peinemann K-V 2006 Hybrid membrane materials with different metal–organic frameworks (MOFs) for gas separation *Desalination* **200** 424-6

IOP Conf. Series: Materials Science and Engineering **839** (2020) 012002 doi:10.1088/1757-899X/839/1/012002

Hardness optimization of epoxy filled PTFE with/without TiO_2 filler

Nur Hidayah Ibrahim, Dr. Ahmad Zafir Romli and Dr Nik Noor Idayu Nik Ibrahim

Facuty of Applied Sciences, Universiti Teknolgi MARA, 40450 Shah Alam, Selangor, Malaysia, Centre of Polymer Composite Research and Technology (PoCResT), Institute of Science, Universiti Teknologi MARA Shah Alam, Selangor, Malaysia

Abstract. Osteoporosis is one of the bone diseases that have the highest number of patients. Throughout the years, there are multiple choices of material that has been used for bone implant such as metals or plastics. The uses of these materials in medical implant are not permanent and they needed to be changed because of its low hardness that would cause pain. In this paper, a new micro bearing concept system by using polymer composite was introduced where epoxy was used as the matrix to improve hardness of micro filler and this composite later can be used in medical implant. PTFE powder filler was used and incorporated at 1%, 3%, 5%, 7% and 10% in the epoxy system respectively. Acetone was used as diluents in the epoxy to improve filler dispersion. Another set of epoxy filled PTFE at 1%, 3%, 5%, 7% and 10% was added with 2% of titanium oxide (TiO_2) and results were compared in terms of hardness and density. Result showed composite with TiO_2 incorporation obtained highest values which were 5% increase at 3%PTFE loading and 1% increase at 10%PTFE loading for hardness and density respectively compared to the composite without TiO_2.

1. Introduction

Hardness is one of the most important factors when it comes to medical implant so that it helps to prolong the life of the implant itself. Problem would arise when the material used for medical implant such as ultrahigh molecular weight polyethylene (UHMWPE), polymethylmetacrylate (PMMA), polyethyletherketones (PEEK), and polytetrafluoroethylene (PTFE) has low hardness [1]. Problem such as osteolysis will arise when the materials have low hardness. Osteolysis is the process where the tissue around the implant is disappearing caused by the articulating surface. The movement of this implant will result in the formation of debris when the materials are soft. Therefore, the implant will not function well and cause pain to the patients [2]. The use of UHMWPE gives excellence result for total joint arthroplasties in long term and PTFE on the other hand has been used as solid lubricant bearing implant in knee implant. PTFE has outstanding properties for this application as it has superior wear resistance. Apart from superior wear resistance, it also gives high fracture toughness. These properties made PTFE to be used broadly in many applications substituting many other materials. PTFE has a very low coefficient when it is rubbed between against two metallic surfaces making it suitable for bearing application [3]. It has lower friction coefficient compared to other engineering materials such as nylon, high density polyethylene (HDPE), low density polyethylene (LDPE), and polyetheretherketone (PEEK). However, the uses of these materials in medical implant are not permanent and they needed to be changed over the years. These materials have low hardness leading to possibility to fractured or dented causing pain to patient.

In this study, PTFE will be used as filler into the epoxy. In some studies, it is proved that adding PTFE filler can potentially enhance wear of epoxy while improving the PTFE hardness. Good mechanical strength and hardness, as well as thermal and chemical resistance, helps the application of epoxy resins as commercially used engineering materials. The behaviour of epoxy material in terms of wear and friction will be improved when the adhesion to the counterpart material is reduced while enhancing its hardness and stiffness. This can be achieved by adding PTFE as filler into epoxy systems [4]. In other studies, the performance of epoxy is modified by adding different fillers. Researchers have found that adding fillers such as alumina (Al2O3) and boron carbide (B4C) will improve mechanical properties of epoxy. As to reduce the friction and the wear environment of epoxy, solid lubricants such as graphite, PTFE and molybdenum sulphide are used [5], [6]. For the last decade, fillers such as carbon fibre and graphite have been widely used to improve the mechanical properties of PTFE [7].

This study produced a solid lubricant micro bearing polymer composite based on the epoxy system as the construction material in which later can be substitute for the plastic bearing in medical implant. Polytetrafluoroethylene (PTFE) was incorporated in the epoxy system. On the other hand, Titanium dioxide was also added into other set of formulations as wear resistance filler and result showed those composite with TiO_2 filler have improved hardness.

2. Experimental details

2.1. Materials and manufacturing

2.1.1. Polytetrafluroethylene powder.
PTFE powder obtained was supplied from GoodFellow Cambridge Limited Huntington PE29 6WR England. It was manufactured from Sigma-Aldrich GF20091743-1EA United Kingdom. The average particle size of the powder was 20 micron. Apart from average particle size, its density was 2.2 g/cm^3. PTFE is non polar material thus it does not dissolve in water or any polar solvent. The powder came in irregular shape and in a large agglomerate size (several hundred microns) rather than fine powders. This condition made it hard to incorporate PTFE into the composite [8]. That led to the need of reducing the size of the clump powder prior to mixing stage. Hence, PTFE powder was blended using steel blender to obtain fine powder. The PTFE powder was then sieved using 360 mesh sieves.

2.1.2. Epoxy and hardener.
The epoxy and hardener (amine) used in this experiment was supplied from Vistec Technology Services and manufactured from Oriental Option SDN BHD. This epoxy and hardener were EPOXEN CP362 part A and EPOXEN CP362 part B type resin and hardener. It has transparent colour and provides glossy hard surface after fully cured. The viscosity of epoxy and hardener were 7000 MPa.s and 300 MPa.s respectively. The mixing ratio for epoxy and hardener was 100 to 50 parts by weight. It took 3 hours to cure and 10 hours to fully cure. Cured epoxy has hardness of 90D and water absorption $<0.5\%$.

2.1.3. Titanium dioxide (TiO_2).
TiO_2 powder (224227) used was manufactured from SIGMA-ALDRICH 3050 Spruce Street, Saint Louis, MO 63103, USA. It came in powder in white colored. The average particle size of TiO_2 was $\leq$ 5.0 micron. The density of TiO_2 powder was 4.23 g/cm^3. Before TiO_2 powder was used, it was also sieved using 360 mesh sieves because some of the powders were also agglomerated. It was used as a filler to enhance the wear properties of the micro-bearing composite.

2.1.4. Acetone.
Acetone (C_3H_6O) used was manufactured from Classic Chemicals SDN BHD Temasya Industrial Park, 40150 Shah Alam, Selangor Darul Ehsan, Malaysia. The appearance was clear and bright. The density of the acetone was 0.792 g/cm^3. It was used to reduce the viscosity of the epoxy as well as to ease dispersing of the filler while mixing.

2.1.5. Methodology.
In this paper, PTFE was used as the filler to be incorporated in the epoxy where epoxy acts as matrix. Instead of obtaining homogenous mixture, PTFE on the other hand should sink forming a layer at the bottom of the composite due to PTFE having greater density compared to epoxy. So instead of floating to the surface it sunk in the bottom. According to different addition quantity, two groups of samples having 1%, 3%, 5%, 7% and 10% of PTFE loading have been prepared for each group. One group of the sample is added with constant amount of TiO_2 at 2% loading.

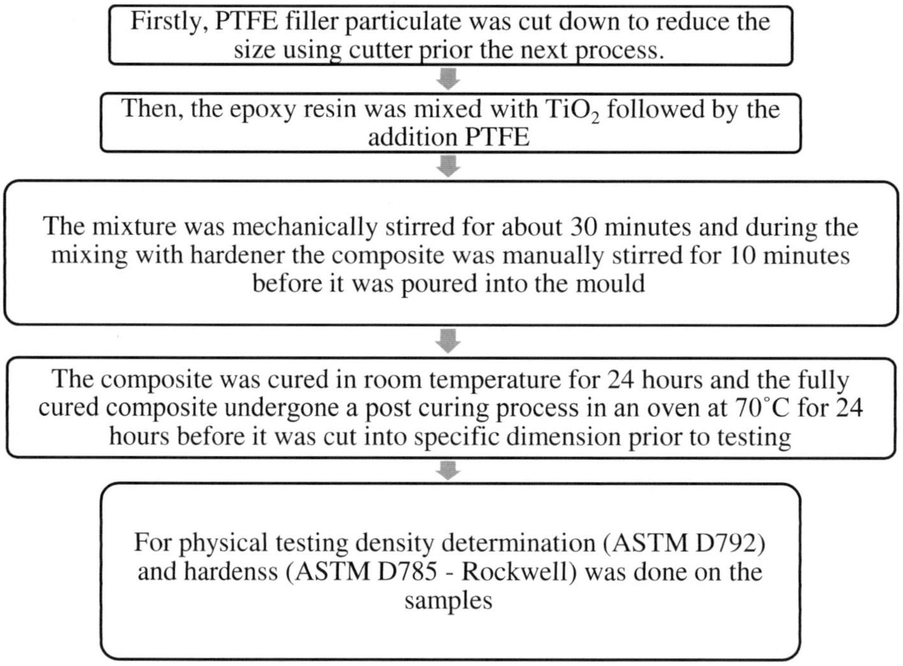

Figure 1. Method flow chart

3. Result and discussion

3.1. Effect of TiO_2 filler on PTFE filled epoxy
Figure 2 and 3 below shows the density and hardness result of the composite. The density test was carried out on specimens by taking 3 samples on average. Density test was done using the Densimeter (g/cm^3). It can be seen on the graphical representing on the figure 2a, the density of the composite was increased with the increased amount of PTFE loading. The increasing filler content increase the volume fractions thus increasing the density value. The properties of composites were significantly influenced by the density property [9].

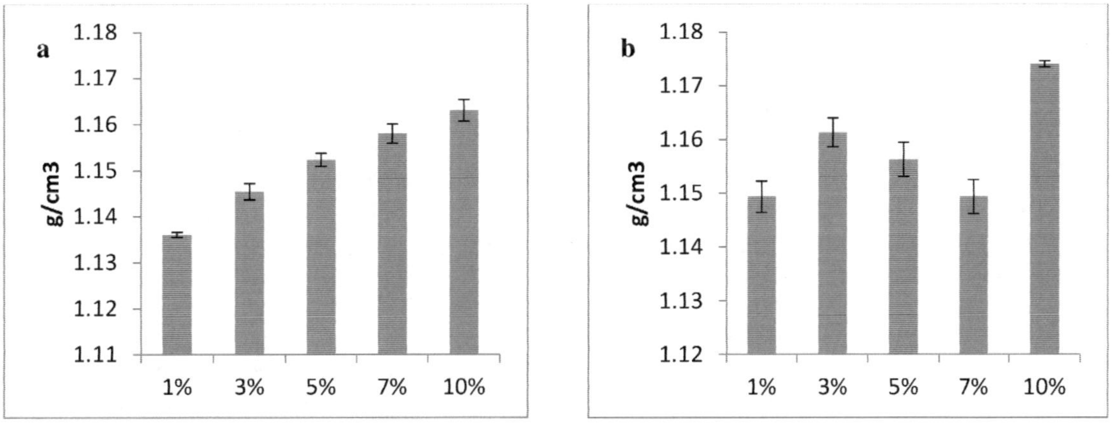

Figure 2. (a) Density without TiO_2 (b) Hardness with TiO_2

The hardness result in figure 3a also showed an increasing trend in the graph. This stated that the hardness of the composites in figure 3a increased as the density in figure 2a increased. There was only 1.4% increase in density for 3% PTFE loading with TiO_2 in figure 2b compared to 3% PTFE loading without TiO_2 filler in filler compared figure 2a. The fluctuation of result can be seen in figure 2b. The main reasons for this are the ceramic powders tend to agglomerate causes bad dispersion in the PTFE matrix which produces more pores in the matrix. It is more difficult for PTFE to have a good adhesion with TiO_2 fillers so that many pores appeared which results in the decline of density. It is difficult for PTFE to have strong adhesion with ceramic powder as PTFE has a low surface energy for 5% and 7% loading fillers as can be seen on figure 2b.

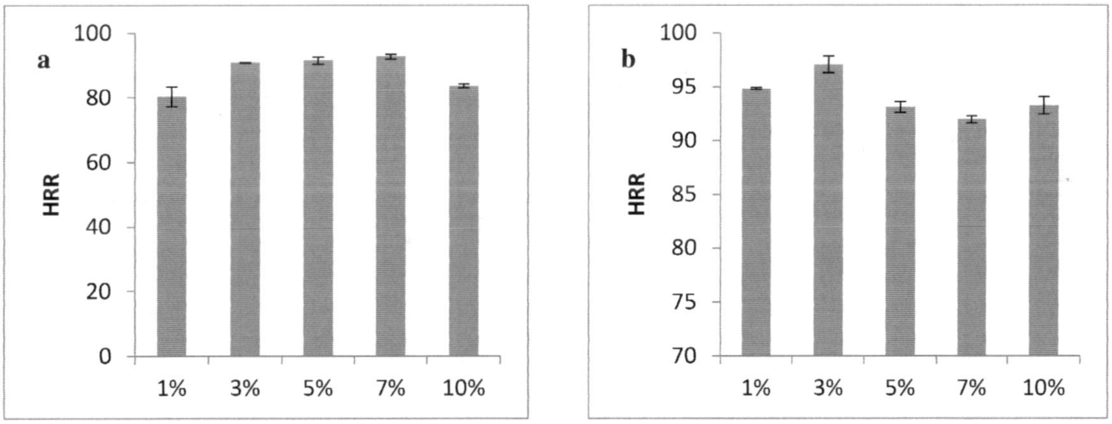

Figure 3. (a) Hardness without TiO_2 (b) Hardness with TiO_2

The hardness test was carried out on specimens by taking 3 sampling points. Hardness test was done using the Rockwell method with an R scale (HRR) which has no unit. The graphical representation of figure 3a showed that the hardness of the composite was slightly increased. The 3% PTFE content was increased at 11.7% than the 1%PTFE loading content followed by the increase of 5% PTFE and 7% PTFE load content at 0.6% and 1.3% respectively when the amount of PTFE filler was added. However, the result showed decreasing at 9.7% in hardness at 10% PTFE loading. Factors such as filler size and shape can greatly affect the hardness of composite [10]. This could be explained when the point where the particles are in contact with the matrix is reached as the result of the increasing of volume-fraction filler. After this point, stress is transferred across the material mostly via (hard) particle-particle interactions [4][10]. This showed the non-linear dependence of modulus, hardness and related properties with filler content, as seen in figure 3a for 1% PTFE to 7% PTFE loading. Hardness is also influenced by the specific composition and structure of the organic matrix.

IOP Conf. Series: Materials Science and Engineering **839** (2020) 012002 doi:10.1088/1757-899X/839/1/012002

This effect is enhanced for certain monomers, but has also been shown to be enhanced in composites with less filler concentration which might be the cause of reduction in hardness value for 10% PTFE loading. As for the graphical representing on figure 3b, it can be seen that the value of hardness generally was higher when 2% of TiO_2 was added in all PTFE loading epoxy composites. The highest hardness could be seen at 3%PTFE loading with TiO_2 incorporation which was 5% higher than the hardness without TiO_2 incorporation. However the decreasing in obtaining the result for composites with added TiO_2 filler at 5% and 7% might be due to problem in term of filler to filler interaction between PTFE and TiO_2 as well as the interaction between filler to matrix of both these two fillers with the epoxy system as discussed at the density section where the ceramic powder tend to agglomerate and PTFE fillers on the other hand has low surface energy causing these particles to have poor adhesion thus producing pores within the matrix making it less compact and decreasing its density as well as hardness value. However, we could see an increase for 10%, this might be due to PTFE filler has become dominant thus contributing in its higher density as well as hardness.

4. Conclusion

The results of testing observations provide few conclusions: the largest value of density, 2%PTFE with TiO_2 incorporation which was 1.4% higher than the composite without TiO_2. The largest hardness value, which was 3%PTFE loading with TiO_2 which was 5% higher than the composite without TiO_2. In conclusion, the TiO_2 filler incorporation increased the density and hardness of the composite.

5. References

[1] Trajanoska K, Schoufour J D, Jonge E A L, Kieboom B C T, Mulder M, Stricker B H, Voortman T, Uitterlinden A G, Oei E A G, Ikram M A, Zillikens M C, Rivadeneira F and Oei L 2018 Fracture incidence and secular trends between 1989 and 2013 in a population based cohort: The Rotterdam Study *Bone* **114** 116–24

[2] Miller M D, Peters C L, and Allen B 2006 Early Aseptic Loosening of a Total Knee Arthroplasty Due to Gore-Tex Particle-Induced Osteolysis *J. Arthroplasty* **21** 765–70

[3] Biswas S K 1992 Friction and wear of PTFE **158** pp 193–211

[4] Zhang Z, Breidt C, Chang L, Haupert F and Friedrich K 2004 Enhancement of the wear resistance of epoxy: Short carbon fibre, graphite, ptfe and nano-tio2 *Compos. Part A Appl. Sci. Manuf* **35** 1385–92

[5] Marat-mendes R, Brandão P, Infante V and Deus A M 2016 An experimental study on mechanical properties of epoxy-matrix composites containing graphite filler modeling of a high pressure turbine blade a , b * of an airplane gas turbine engine *Procedia Struct. Integr* **1** 74–81

[6] Sudheer M, Hemanth K, Raju K and Bhat T 2014 Enhanced Mechanical and Wear Performance of Epoxy / glass Composites with PTW / Graphite Hybrid Fillers *MSPRO* **6** 975–87

[7] Desale D D and Pawar H B 2018 Performance Analysis of Polytetrafluoroethylene as Journal Bearing Performance Analysis of Polytetrafluoroethylene as Journal Bearing Costing models for capacity optimization in Industry 4 . 0 : Trade-off between *Procedia Manuf* **20** 414–19

[8] Ebnesajjad S and Khaladkar P R, *Properties of Neat (Unfilled) and Filled Fluoropolymers*

[9] Luo F, Yuan Y, Tang B, Yang B and Zhang S 2017 The Effects of TiO2 Particle Size on the Properties of PTFE/TiO2 Composite *J. Mater. Sci. Chem. Eng* **05** 53–60

[10] Alshabib A, Silikas N and Watts D C 2019 Hardness and fracture toughness of resin-composite materials with and without fibers *Dent. Mater* **35** 1194–203

Acknowledgement
The authors would like to thanks Universiti Teknologi MARA Malaysia, the Institute of Research Management and Innovative (IRMI) for the assistance throughout the research. This research was funding by Ministry of Education Malaysia (MOE) under the Fundamental Research Grant Scheme (FRGS), 600-IRMI/FRGS 5/3 (114/2019).

Fabrication and magnetic properties of Co based thin films synthesized by physical vapor deposition under vacuum

A Kharmouche

Laboratory of Surfaces and Interfaces Studies of Solid Materials (LESIMS), Faculty of Technology, Ferhat ABBAS Sétif1 University, Sétif, 19000, Algeria
kharmouche_ahmed@univ-setif.dz

Abstract. We used thermal heating process, under a pressure of 10^{-7} mbar, to fabricate thin films of cobalt on monocrystalline silicon Si (100) substrates. The incident beam strikes the substrates under normal and oblique incidences within a home-made evaporation chamber. The thickness of the deposited films ranges from 18 to 400 nm. Microscopic characterizations of the films are performed with X-ray diffraction (XRD) measurements and infer that all the samples are polycrystalline, with an hexagonal close packed (h.c.p.) structure and exhibit a <0001> preferred orientation, for the perpendicularly evaporated films. The grain size is found to increase with the thickness of the magnetic layer. The values of the computed parameters allow concluding to a compressive stress for the thinnest films. The static magnetic properties have been performed by means of magnetic force microscopy (M.F.M.) and Alternating Gradient Field Magnetometer (A.G.F.M.) tools. The dynamic magnetic properties have been investigated using Brillouin Light Scattering (B.L.S.) measurements. The M.F.M. observations were performed after in-plane ac demagnetization, and stripe patterns are only observed for the thickest films, showing the weaker perpendicular anisotropies. We used the experimental results provided by these tools to compute the effective magnetic anisotropy factors K_u. Values of K_u anisotropy constants, higher than 13 $\times$ 10^6 erg.cm^{-3}, have been found.

Keywords: X-ray diffraction, Thin films, Hysteresis, Magnetic anisotropy, Brillouin Light Scattering (B.L.S.).

1. Introduction

Ferromagnetic thin films have been intensively studied in the last two decades due to the development of new and sophisticated elaboration methods and techniques of analysis, particularly Co and alloys such as CoCr and CoPt etc [1-5]. Some materials present a desirable perpendicular magnetic anisotropy and can then be good candidates for high density storage media; their magnetic properties depend greatly on the methods and conditions of preparation and they are affected by the structural properties. In this work we report structural, static and dynamic magnetic studies of perpendicularly and obliquely evaporated Co/Si (100) thin films as a function of Co thickness. To characterize these films, a lot of techniques were performed: X-ray diffraction, Alternating Gradient Field Magnetometer (AGFM), Brillouin Light Scattering (BLS), Magnetic Force Microscopy (MFM).

2. Experimental procedures

We have evaporated, using thermal heating process under vacuum, two series of Co thin films onto monocrystalline Si (100), perpendicularly and obliquely to the targets. The 99.99 % purified cobalt powder was placed in a crucible made of tungsten coated with alumina. Before deposition, the pressure in the evaporation chamber was around 10^{-7} mbar; during the evaporation the base pressure was lower than 2×10^{-6} mbar. Each series is composed of several samples with thickness ranging from 18 to 400 nm. X-ray diffraction measurements were performed using a Siemens D-500 diffractometer with $\lambda = 1.789$ Å. The hysteresis loops of these Co films were carried out by AGFM measurements, with the external magnetic field H applied perpendicular (polar configuration) and in the film plane (longitudinal configuration), and the zero-field magnetic structure has been investigated by Magnetic Force Microscopy (MFM), using a Veeco 3100 apparatus. Finally, the magnetic anisotropy in Co/Si samples was investigated by means of Brillouin Light Scattering (BLS) experiments, using a (2 × 3)–pass tandem Fabry-Perot interferometer. The thin films were illuminated by a single-mode Ar+ ion laser at the wavelength of $\lambda = 5145$ Å, with an incident power of 100 mW. The value of the transferred in-plane wave vector $q_{//}$ is connected to θ by the relation $q_{//} = (4\pi/\lambda) \sin\theta$. The experiments were done with different values of the intensity of the applied magnetic field, in the 0–5 kOe range.

3. Results and discussion

3.1. Structural study
The XRD spectra recorded for perpendicularly evaporated Co/Si samples exhibit 3 Bragg peaks, <10-10>, <0002> and <10-11> of hexagonal close packed (h.c.p.) Co, the most intense one being <0002> peak. All the samples are polycrystalline, with a h.c.p. structure and exhibit <0001> preferred orientation. Using the diffraction spectra we computed the values of the lattice constants a and c. These values are 2.49 Å and 4.69 Å, respectively. The bulk parameters being $a = 2.51$ Å and $c = 4.07$ Å, our Co films are therefore under a tensile stress due probably to the growth mode of the films, a Stranski-Krastanov growing mode.

3.2. Static magnetic study
3.2.1. *Hysteresis curves.*
Figure 1 displays examples of hysteresis curves obtained for the longitudinal configuration (H parallel to the film plane) and the polar configuration (H perpendicular to the film plane). In the longitudinal configuration, it is easy to saturate the films, whereas for the perpendicular configuration, we have to apply a very high magnetic field to come within reach of saturation and this is typical to in plane magnetic anisotropy. From the shape of the curves we infer that the easy magnetization axis lies in the film plane, for all samples.

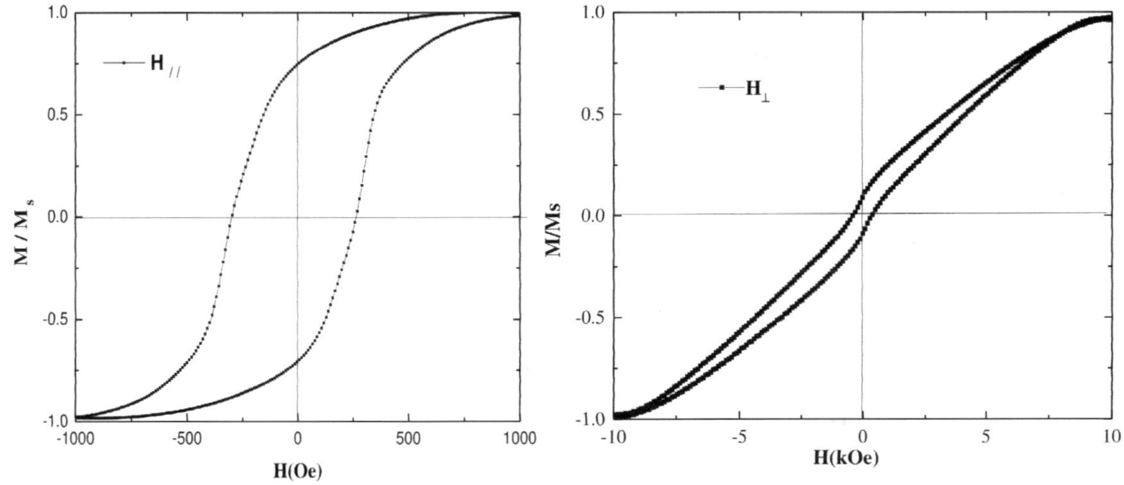

Figure 1. Examples of hysteresis curves obtained for the longitudinal configuration (H parallel to the film plane) and the polar configuration (H perpendicular to the film plane).

3.2.2. *MFM observations.*

With the aim to examine the domain configurations of our evaporated Co thin films, we used the magnetic force microscope to observe the images. Figure 2 is an image of the surface topography (a) and stripes domains (b) recorded for the thicker sample, Co (400 nm) thin film. The films present very smooth surface topography, the rms roughness being equal to a few angströms. Obtained with a lift scan of 50 nm, the MFM image (b) shows clearly the presence of magnetic domains; the half period of these weak stripe patterns is equal to 200 nm. These weak stripes patterns are evidence of the existence of an out-of-plane component of the magnetization vector M in the thickest samples in which the magnetocrystalline anisotropy is dominant. The light and dark areas correspond to the site of up and down domains.

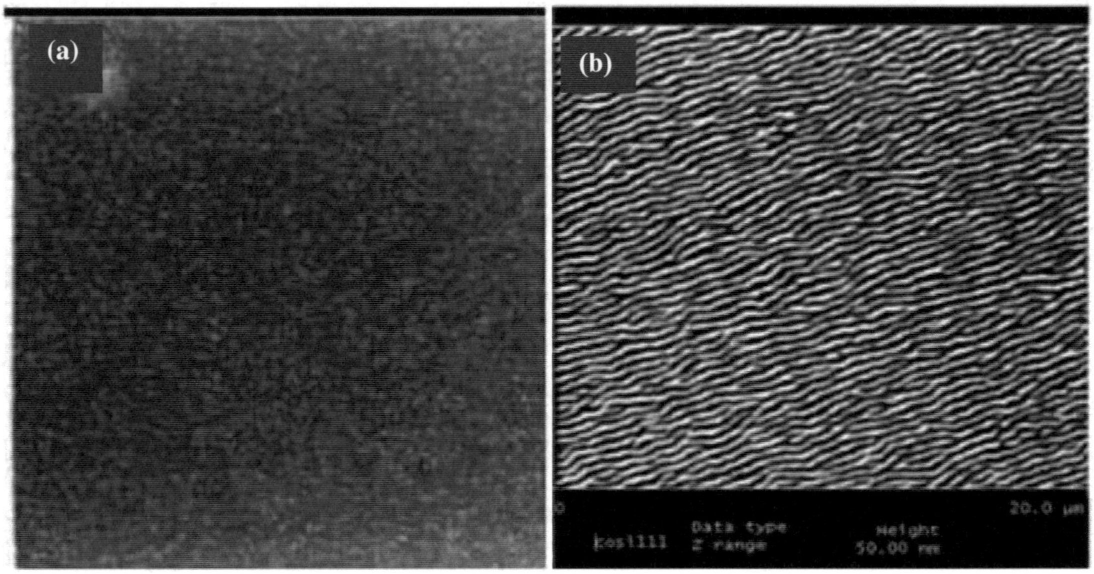

Figure 2. AFM (a) and MFM (b) images of Co (400 nm) thin film. Scan area is 20 × 20 µm.

3.3 Dynamic magnetic study

3.3.1. Brillouin light scattering.

In the most recent decades, Brillouin Light Scattering technique has been used to explore magnetic materials properties [6–9] and precisely the magnetic anisotropy.

BLS experiments were performed on our thin films under the conditions described in Section 2. In Figure 3, we show specimens of experimental and calculated Brillouin spectra obtained with H = 5 kOe for Co (195nm)/Si sample. As can be seen on Figure 3, the agreement between the experimental and the calculated spectra is good enough. It was obtained by fitting the magneto-crystalline anisotropy field H_a (assumed to be uniaxial, $H_a = 2K_u/M_s$, with K_u the uniaxial anisotropy constant).

In Figure 4, we display typical frequency shift as a function of applied magnetic field H curves for two representative samples Co (195 nm)/Si and Co (50 nm)/Si films. The experimental data for the Damon-Eshbach (DE) are the points and the simulated predictions are the solid lines; here too, the fit is very satisfactory.

In Table 1, we present the computed magnetic anisotropy factors as a function of the thickness t of the magnetic layer, for the Co/Si thin films, evaporated either perpendicular to the substrate or with an oblique incidence angle.

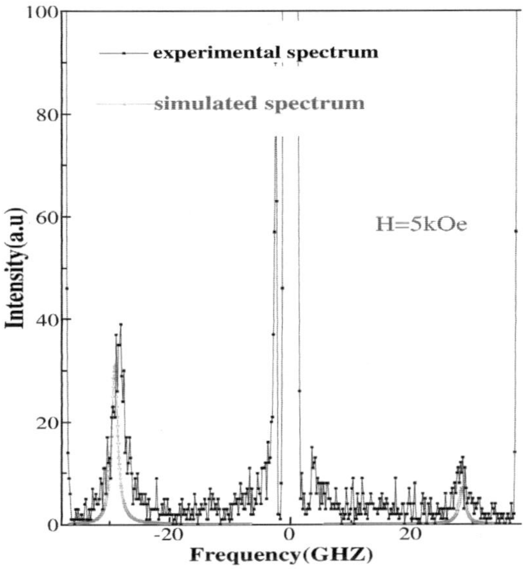

Figure 3. Experimental and calculated B.L.S spectra of Co (195 nm)/Si film.

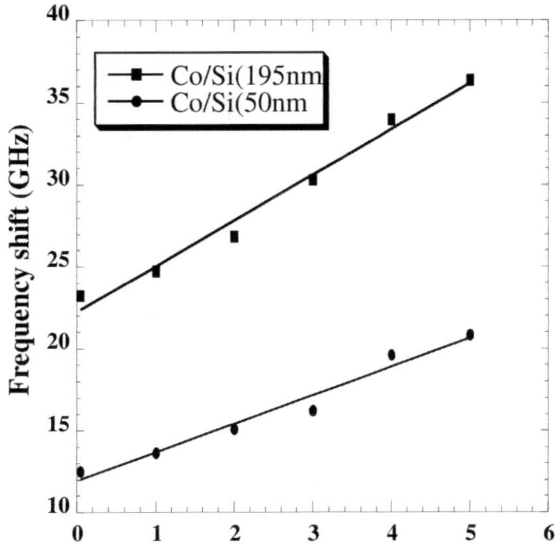

Figure 4. Frequency shift versus applied magnetic field H, for Co (195nm) / Si and Co(50nm) /Si. Experimental data (points), theoretical predictions (solid line).

Table1. Anisotropy field versus thickness, for perpendicularly and obliquely evaporated films.

t(nm)		18	50	195	280	400
K_u (10^6 erg.cm^{-3})	oblique	13.7	12.6	8.4	8.1	7.9
	perpendicular	8.8	7.6	5.2	4.3	3.1

As can be seen in Table 1, K_u decreases drastically, from 8.8 to 3.1 × 10^6 erg.cm^{-3} for the films perpendicularly evaporated, and decreases rapidly too, from 13.7 to 7.9 × 10^6 erg.cm^{-3} for the films obliquely evaporated, as the thickness of the magnetic layer increases from 18 to 400 nm.

A number of phenomena are at the origin of the magnetic anisotropy in thin films, namely the magnetocrystalline structure and the two-dimensional shape of the material. In the Co (18 nm) thin film, the demagnetizing field is important, which makes the shape anisotropy preponderant, and consequently, gives rise to a high value of *Ku*. In the Co (400 nm) thin film, the demagnetizing field is lower which makes the shape anisotropy lesser and infers a relatively low value of K_u, due principally to the magnetocrystalline anisotropy.

4. Conclusion

With a homemade evaporation chamber we synthesized series of Co thin films, deposited either perpendicular or with an oblique incidence angle with regards to the Si (100) substrate. The hysteresis curves display loops inferring that the easy magnetic axis is lying in the plane of the film. Well-defined stripes patterns are observed in the thicker films. From the Brillouin Light Scattering study we computed the magnetic anisotropy factors and found values as high as 13×10^6 $erg.cm^{-3}$ for the obliquely evaporated films, much greater than those found for perpendicularly evaporated samples.

5. References

[1] Nozawa N, Saito S, Hinata S and Takahashi M 2013 Large uniaxial magnetocrystalline anisotropy for Co50Pt50 disordered alloy films with hexagonal-close-packed stacking structure by substituting Pt with Rh *J. Phys. D: Appl. Phys.* ***46***(17) 172001

[2] Melloul A and Kharmouche A 2019 Synthesis, structure and magnetic properties of Co_xFe_{100-x} thin films thermally evaporated onto Si (111) substrate *J. Mater. Sci: Mater. Electron.* **30** 13144-13150

[3] Hu H, Chen H, Yu S, Chen L, Chen J and Wu G 2006 Fabrication and magnetic properties of Co_x Pd_{1-x} composite nanowire *J. Magn. Magn. Mater.* **299** 170–175

[4] Kharmouche A, Cherif S, Bourzami A, Layadi A and Schmerber G 2004 Structural and magnetic properties of evaporated Co/Si(100) and Co/glass thin films *J. Phys. D: Appl. Phys.* **37** 2583

[5] Kharmouche A 2013 Magnetic anisotropy factors of vapor deposited CoCr thin films on Si and glass substrates *J. Magn. Magn. Mater.* **327** 91–94

[6] Hillebrands B, Mathieu C, Bauer M, Demokritov S, Bartenlian B, Chappert C, Decanini D, Rousseaux F and Carcenac F 1997 Brillouin light scattering investigations of structured permalloy films *J. Appl. Phys.* **81** 4993

[7] Büttner O, Bauer M, Demokritov S, Hillebrands B, Kivshar Y S, Grimalsky V, Rapoport Y and Slavin A N 2000 Linear and nonlinear diffraction of dipolar spin waves in yttrium iron garnet films observed by space- and time-resolved Brillouin light scattering *Phys. Rev. B* **61** 11576

[8] Serga A, Schneider T, Hillebrands B, Demokritov S and Kostylev M 2006 Phase-sensitive Brillouin light scattering spectroscopy from spin wave packets *Appl. Phys. Lett.* **89** 063506.

[9] Sebastian T, Schultheiss K, Obry B, Hillebrands B and Schultheiss H 2015 Micro-focused Brillouin light scattering: imaging spin waves at the nanoscale *Front. Phys.* **3** 35

Acknowledgement

This work is funded under the grant of the Algerian General Directorate for Scientific Research and Technological Development (DGRSDT).

Sonochemical synthesis of gold nanoparticles via palm oil fronds extracts for cytotoxicity assay

Adamu Ibrahim Usman*[1,3], Azlan Abdul Aziz[2] and Pegah, Moradi Khaniabadi[1]

[1]School of Physics University Sains Malaysia, 11800, Pulau Pinang Malaysia,
[2]Institutes for Research in Molecular Medicine, USM, 11800, Pulau Pinang Malaysia,
[3]Department of Physics, Federal University Kashere, 771103, Gombe State Nigeria
*izelibrahim@rocketmail.com

Abstract. One of the major focus of scientific community todays is the synthesis of nanoparticles using various bio agents, which are biodegradable for their subsequent bio-application. One pot green synthesis of gold nanoparticles (AuNPs) was successfully achieved using palm oil fronds extracts as stabilizing agent and ultrasound radiation as reducing agent. The functional groups such as OH^-, CO – stretch, NH – stretching and other function groups were found to be responsible for stabilizing the AuNPs. The size and morphology of the as-synthesized AuNPs was investigated using HRTEM. The XRD patterns display the peaks of face centred cubic crystal structures at (111), (200), (220), (311) and (222). The cytotoxicity of the bio-synthesized was investigated using human embryonic kidney 293 cells (HEK-293 cells) and breast cancer cell (MCF-7 cells) in vitro study, where more than 50 % of the cells were viable after exposure from 0 mg/ml (negative control) to 1 mg/ml, which indicates the material suitability in future medical application.

Keywords: gold, nanoparticls, cytotoxicity, cells, synthesis

1. Introduction

The engineering of nanoparticles (NPs) has been broadly advanced for the past decades, numerous NPs with distinct chemical and physical properties have been developed for various medical applications [1,2]. From our lab various nanoparticles have been synthesized using sonochemical method and have demonstrated impressive results [3–7]. The interaction between nanoparticles and biological cells required improvement especially at cellular level, where the biological risk is optimum for effective evaluation and control for biomedical application [8]. Elucidating the interactions between biological system and the physiochemical properties of NPs responses, is vital [9]. Such studies are requisites for producing and manipulating NPs for accessing their toxicity and responses on biological system. The recognition of the Gold nanoparticles (AuNPs) as potential nanoprobes in clinical transformation for biomedical application is one of the major concern of the scientific community, and they were once considered to be bio-friendly but results to unanticipated toxicity and cell cycle manipulation to mammalian cells [10]. Even though, AuNPs have been recognized as promising nanoprobes in biomedical applications for clinical translation, however the results of various studies have differed due to the usage of various AuNPs synthesized with different physicochemical properties [11]. Thus, further studies are required to evaluate the mechanisms through which the toxicity of the AuNPs affect the cell cycle due to their exposure in an invitro system.

In this article, green AuNPs coated with bio agent (POFE) was synthesized via sonochemical method to evaluate their cytotoxicity on HEK-293 and MCF-7 cells for suitable medical application.

2. Methodology

Hydrogen tetrachloroaurate acid ($HAuCl_4.3H_2O$) was procured from Sigma-Aldrich Corporate (St. Louis, MO, USA) and used without alteration. Double distilled water (DDW) (at 25°C, conductivity of 10^{-6} Sm^{-1} and surface tension of 72.0 mNm^{-1}) was used throughout the experiment as solvent. Palm oil fronds (POF) were obtained from Palmco Holding Sdn. Bhd. Penang, Malaysia and prepared according to [3]. All the consumable and cell proliferation reagent (WST-1) were obtained from Modern-Lab Chemical Sdn. Bhd. Penang, Malaysia. While human embryonic kidney cells (HEK-293) and breast cancer cell (MCF-7) were obtained from American Type Culture Collection (ATCC, Rockville, MD, USA). 10 ml of the extracts from POFs was put into a beaker containing 100 ml of 0.01% (w/w) of the gold precursor. The reaction was expose to ultrasound radiation for 10 minutes. The formation of the AuNPs were observed within the first two minutes due change in colour. The average hydrodynamical and grain size of the bio-synthesized AuNPs were characterized using dynamic light scattering (DLS) and high resolution transmission electron microscopy (HRTEM) respectively. The functional group responsible for stabilizing the nanoparticles (NPs) were identified using spectroscopy Perkin Elmer System 2000 and the UV-vis spectral for Surface Plasmon Resonance (SPR) measurements of the AuNPs using Shimadzu in the range of 400–700 nm. The crystallinity of the AuNPs was determined using D8 ADVANCE BRUKER with Cu-radiation of λ = 0.154 Å regulated at the current 40mA and voltage of 40 kv. The cytotoxicity of the synthesized AuNPs was access using cell proliferation reagent WST1 assay. The WST1 assay study adopted here was based on Rosli and co. [12] with modification. The cells were culture in a culture flask 25 cm^2 with Dulbecco's Modified Eagle's Medium (DMEM) and incubated in humidified atmosphere incubator at 37°C and % CO_2 supply. When the cells have reached 80% to 90% confluence and ready to be treated, the DMEM was removed and the cells were washed with 3 ml of sterile Phosphate Buffered Saline (PBS) 2X. The trypsin of 400 ml was added to detach the cells from the flacks. An approximate of $1x10^5$ cells ml^{-1} of HEK-293 and MCF-7 were seeded in a flat-bottomed 96-well plate (Figure 1). The cells were incubated for 1 day under 37°C and 5% CO_2 supply. Various concentration of the synthesized AuNPs using POFE as stabilizing agents (10, 30, 50, 70, 90, 150, 250, 500, and 1000 μg/ml) were added to the plates. The cells were then incubated for 24 hours and 10 μl of WST-1 reagent was added to each well and incubated again for 4 hours till Formazan crystals is formed in each well. The plate was immediately read by enzyme-linked immunosorbent assay (ELISA) at 450 nm with reference wavelength of 620 nm. The viability [optical density (OD)] of the cells were calculated using the following expression [13].

$$cells\ viability = \frac{absorbance\ of\ treated\ cells}{absorbance\ of\ untreated\ cells} X\ 100\% \quad (1)$$

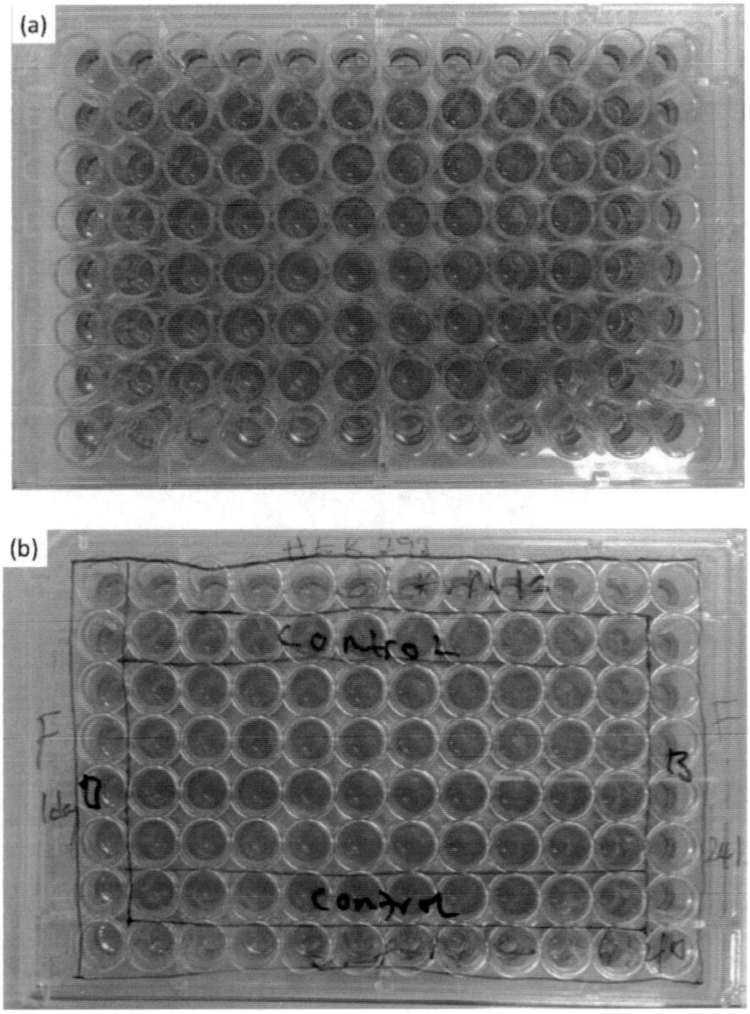

Figure 1. Seeded HEK293 cells in a flat-bottomed 96-well plates (a) without gold nanoparticles (b) with gold nanoparticles.

3. Results and Discursion

3.1 Surface Plasmon Resonance (SPR) and Crystallinity analysis

The appearance of the ruby red due to subsequent reduction of Au^{3+} from $HAuCl_4.H_2O$ with palm oil fronds extracts (POFE) confirmed the formation of AuNPs. Also, the manifestation of the peak at 542.5 nm after the synthesis from the UV-vis spectral analysis confirmed the formation of the AuNPs (Figure 2a). The induction of the ruby colour is due to collective oscillation of the nanoparticles by electromagnetic field [14]. the crystallinity of the synthesized AuNPs was successfully determined as shown in the XRD diffraction with the following millers indices (111), (200), (220), (311) and (222) corresponding to the reflection peaks emerged at 38.2°, 44.4°, 64.6°, 77.5° and 81.7° respectively (Figure 2b). These features revealed the characteristics of face-centered cubic (FCC) structure. Similarly, The broadening of the peaks confirmed the formation of the NPs [15].

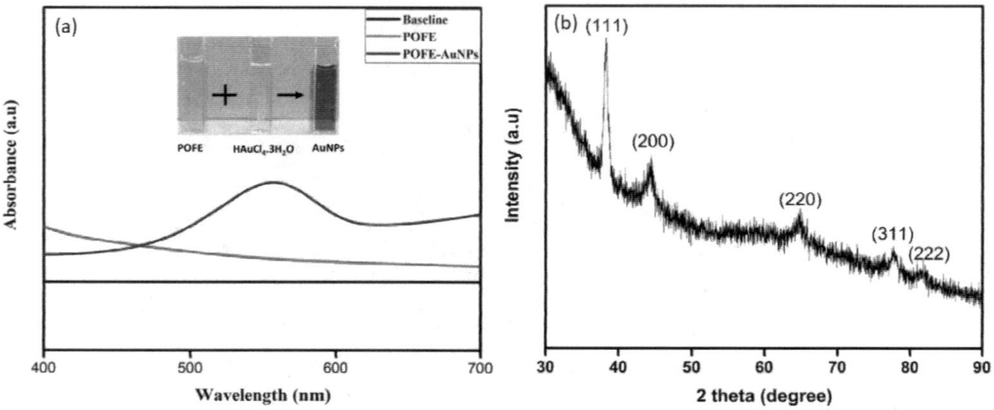

Figure 2. (a) UV-vis spectrum of AuNPs synthesized using POFE and (b) Crystallinity nature of the dried AuNPs synthesized using POFE.

3.2 Morphological and stability analysis

The morphology and the average grain size of the synthesized AuNPs displayed by HRTEM are monodisperse, which are mostly spherical with some projection of triangular and hexagonal shapes with an average grain size of 23 nm (Figure 3a), while, an average of 42.65 nm of the hydrodynamical size was obtained via DLS. The FTIR spectral results for both POFs extracts and synthesized AuNPs indicates presence of phenol functional group (OH^-) and secondary amide (NH^-), which are responsible for stabilization of the synthesized AuNPs and the peak shifts from higher peak (3325 cm^{-1}) before the synthesis to the lower peak (3318 cm^{-1}) after the synthesis (Figure3b) indicates the formation of AuNPs [16].

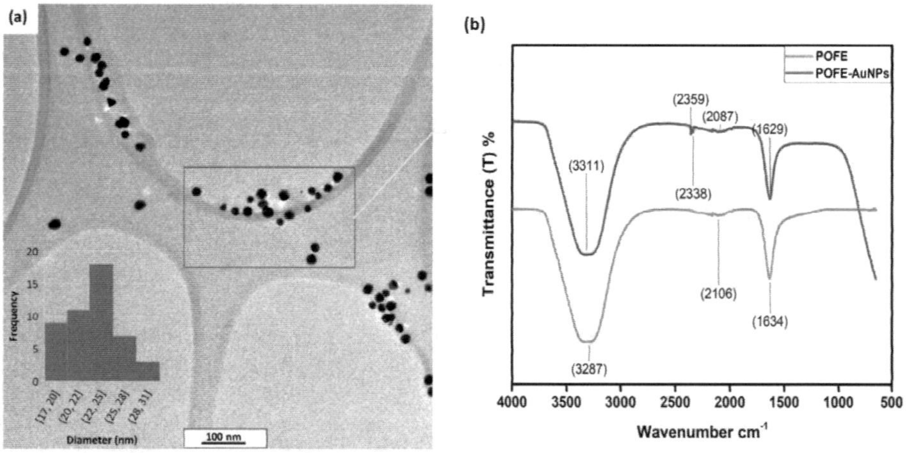

Figure 3. (a) HRTEM analysis (b) FTIR analysis of the AuNPs synthesis using POFE as stabilizing.

3.3 Cytotoxicity studies

A conflicting studies do exist on the cytotoxicity of AuNPs [17]; However, it is believed that size plays a key role in cytotoxicity of AuNPs [18]. The cytotoxicity of the biosynthesized AuNPs using POFE as stabilizing agents was investigated using WST-1 assay on HEK-293 cells and FCM-7 cells at ten different concentrations(10, 30, 50, 70, 90, 120, 150, 250, 500, and 1000 μg/mL). The inhibitory concentration (50% viability or IC_{50}) of the FCM-7 cells after 24 hours of incubation showed more than 95% viability level for 1 mg/mL while about 75% viability was observed when using HEK-293 cells. However, when the incubation time was extended to 72 hours there was decreased in the cell viability with inhibitory concentration value of IC_{50} = 739.53 µg/mL (Figure 4a) and IC_{50} = 872.23 µg/mL (Figure 4b) for HEK-293 and MCF-7 cells respectively. The reduction in the viability of HEK-293 cells might be due to stress caused to the cells because the media nutrient level have been depleting and altered by the addition of the AuNPs concentration and of course the HEK-293 are very sensitive to any change in the media and are not strong to resist the change in the media as FCM-7 do [18]. Based on literatures on the half maximal inhibitory concentration (IC_{50}) of AuNPs [19], it can be concluded that the biosynthesized AuNPs using POFE as surfactants is suitable for future medical application.

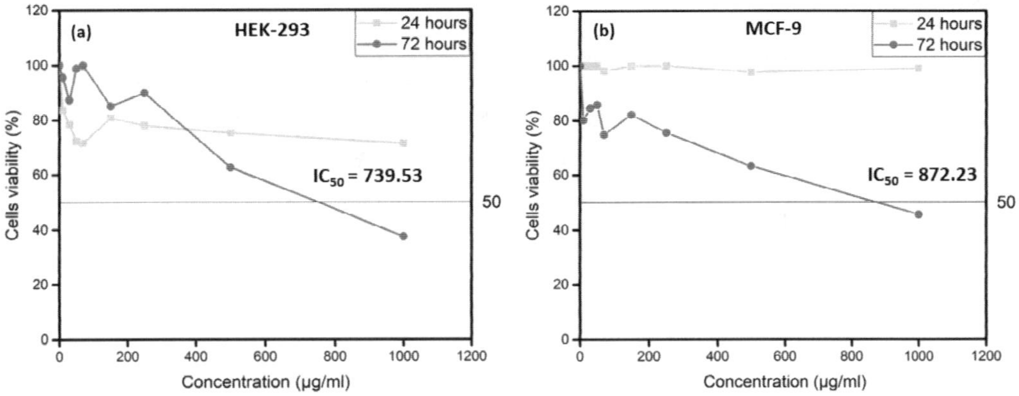

Figure 4. Cytotoxicity for different concentration of the AuNPs synthesized using POFE as stabilizing agent on (a) HEK293 and (b) for 24 hours and 72 hours.

4. Conclusion

In this work, gold nanoparticles (AuNPs) were successfully synthesized using biodegradable stabilizing agents (POFE) via sonochemical method. The synthesized AuNPs revealed crystalline nature base on XRD results. The hydrodynamical size and grain size were obtained to be 42.65 nm 23 nm respectively. HO^- and NH^- were responsible for stabilizing the synthesized AuNPs, the inhibitory concentration obtained after 72 hours incubation in a media containing AuNPs at the value of IC_{50} = 739.53 µg/mL and IC_{50} = 872.23 µg/mL for HEK-293 and MCF-7 cells respectively indicates the suitability of the synthesized AuNPs for future medical application.

5. References

[1] P. Huang, B.; S.Yan; L. Xiao; R. Ji; L.Yang; L. Miao, A.J. Wang 2018 Label-free imaging of nanoparticle uptake competition in single cells by hyperspectral stimulated raman scattering, *Small*, **14** p. 1703246,.

[2] B. Maximilien, J., Beyazit, S., Rossi, C., Haupt, 2015 Nanoparticles in Biomedical Applications. In Measuring Biological Impacts of Nanomaterials, *Springer Cham, Switz.*, pp. 177–210.

[3] A. I. Usman, A. A. Aziz, and O. A. Noqta 2018 Bio-synthesis of triangular and hexagonal gold nanoparticles using palm oil fronds' extracts at room temperature, *Mater. Res. Express*, **5** 1.

[4] N. A. Hammed, A. A. Aziz, A. I. Usman, and M. A. Qaeed 2018 "The sonochemical synthesis of vertically aligned ZnO nanorods and their UV photodetection properties: Effect of ZnO buffer layer," *Ultrason. Sonochem.*

[5] A. I. Usman, A. A. Aziz, and O. A. Noqta 2019 Green sonochemical synthesis of gold nanoparticles using palm oil leaves extracts, *Mater. Today Proc.* **7** pp. 803–807.

[6] A. I. Usman, A. A. Aziz, and O. A. Noqta, Bio-synthesis of triangular and hexagonal gold

nanoparticles using palm oil fronds' extracts at room temperature, *Mater. Res. Express*, **5**.
[7] A. I. Usman and A. Abdul Aziz, 2018 Optimization of biosynthesis gold nanoparticles via central composite design toward monodisperse, *Mater. Res. Express*, **6** 1, p. 01505.
[8] M. A. Wani, M.Y.; Hashim, M.A.; Nabi, F.; Malik, "Nanotoxicity: Dimensional and morphological concerns," *Adv. Phys. Org. Chem.*, pp. 1687–1702, 2011.
[9] S. El-Ansary, A.; Al-Daihan 2009 "The toxicity of therapeutically used nanoparticles an overview," *J. Toxicol*, p. 754810.
[10] R. H. and J. Q. Qian Li, Chun Huang, Liwei Liu 2018 Effect of Surface Coating of Gold Nanoparticles on Cytotoxicity and Cell Cycle Progression, *nanomaterials*, pp. 1–13.
[11] S. J. Soenen, W. J. Parak, J. Rejman, and B. Manshian 2015 (Intra) Cellular Stability of Inorganic Nanoparticles: Effects on Cytotoxicity, Particle Functionality, and Biomedical Applications, *Chemical reviews*, **115** pp. 2109-2135.
[12] N. S. binti Rosli, A. A. Rahman, A. A. Aziz, and S. Shamsuddin 2015 Determining the size and concentration dependence of gold nanoparticles in vitro cytotoxicity (IC50) test using WST-1 assay, in *AIP Conference Proceedings*, **1657**, no. 1, p. 060001.
[13] N. Parvin and T. K. Mandal 2016 Synthesis of Highly Fluorescence Nitrogen Doped Carbon Quantum Dots Bioimaging Probe, Their In vivo Clearance and Printing Applications **6, p** 18134-18140.
[14] M. Grzelczak, J. Pe, P. Mulvaney, and L. M. Liz-Marza 2008 Shape control in gold nanoparticle synthesisw, *Chem. Soc. Rev.*, vol. 37, pp. 1783–1791.
[15] K. B. Narayanan and N. Sakthivel 2008 Coriander leaf mediated biosynthesis of gold nanoparticles, *Mater. Lett.*, vol. 62, no. 30, pp. 4588–4590.
[16] P. Bollella *et al.* 2017 Green Synthesis and Characterization of Gold and Silver Nanoparticles and their Application for Development of a Third Generation Lactose Biosensor, *Electroanalysis* **29** 1, pp. 77–8.
[17] A. M. Ostdiek, S. Grant, and D. Grant 2015 Mechanical and in vitro characterisation of decellularised porcine aortic tissue conjugated with gold nanoparticles as a vascular repair material **6** 1, pp 1-17.
18] Y. Pan *et al.* 2007 Size-Dependent Cytotoxicity of Gold Nanoparticles, *Small*, **3** pp. 1941–1949.
[19] M. F. H. C. & F. B. Jr. 2016 Gold nanoparticles: A critical review of therapeutic applications and toxicological aspects, *J. Toxicol. Environ. Heal. Part B*, **19** 3–4, pp. 129–148.

Acknowledgments
We want to acknowledge the financial and facilities assistance provided by TETFUND, Abuja Nigeria; Federal University of Kashere, Gombe state Nigeria; and Universiti Sains Malaysia for making this work possible.

Stress-strain behaviour and strength properties of ambient cured geo-polymer concrete

M Talha Junaid, Abdul Saboor Karzad and Salah Altoubat

Department of Civil and Environmental Engineering, Sustainable Construction Materials and Structural System Research Group, University of Sharjah, Sharjah, UAE

Abstract. Given the high environmental impact of OPC concrete on CO_2 emissions, the construction industry is moving towards low carbon alternate binders. In recent years geo-polymer concrete (GPC) has come to the forefront of such efforts. However, challenges remain for GPC to be considered as a main-stream construction material. Since GPC is relatively new material in construction industry, many of its mechanical properties are still not fully understood. The current work looks at the strength properties of ambient cured GPC. This research paper is part of an ongoing investigation on GPC and reports some results of the mechanical properties of ambient-cured GPC. In this study, the stress-strain behavior and compressive strength of such GPC and its dependency on various parameters (ratio and type of activator and proportions & type of pre-cursor materials) is investigated. The results indicate that the compressive strength of GPC is greatly dependent on the proportion of Ground Granulated Blast Furnace Slag (GGBFS) and ratio of activators. It was found that increasing the percentage of GGBFS increased the compressive strength but at the expense of setting time of the mix. The mixes that had GGBFS around 20% of the binder material returned a compressive strength of less than 25 MPa which may not be suitable for structural applications. Whereas, the increase in percentage of GGBFS from 20 to 80% resulted in compressive strength of 40 MPa and higher. The stress-strain relationship of the mixes was compared to two analytical models existed in the literature.

1. Introduction

In the 21st century the concrete industry is facing a challenge to meet the needs of vast infrastructures and housing demand as a result of urbanization and industrialization. At the moment, Portland cement is the favourite option of binder for modern construction industry [1]. However, it is well known that producing Portland cement requires high energy and the CO_2 emissions in this process are substantial. These are the major factors contributing to global warming [1, 2]. Global warming to which cement production also contribute, is considered a life-threatening issue that needs to be tackled. On the other hand, there is a need for disposal of millions of tonnes of industrial by products. To solve both these issues one can incorporate industrial by products into concrete production as supplementary cementitious materials (SCMs) [1].

Currently after water concrete is the second widely used material with cement being its main ingredient [3]. The global production of OPC is steadily increasing without sign of slowing down [4], [5]. In the last twenty years' extensive efforts has been concentrated by the concrete scientific community to decrease the CO_2 emissions by searching alternatives to OPC concrete. Being one of the alternatives, GPC concrete sometimes called green concrete have come to the forefront [6, 7]. GPC can be formed by reacting metallic alkaline (such as sodium/potassium hydro-oxide and/or sodium silicate) with materials containing considerable amount of silica (SiO_2) and alumina (Al_2O_3) such as fly ash (FA), GGBFS and rice husk ash [2, 8]. Such binders when mixed with traditional aggregates produce GPC which can be engineered with properties similar to or sometimes superior to OPC concrete. Moreover and in contrast to OPC, GPC does not rely on calcining $CaCO_3$, which is the main source of CO_2 emissions [9–11].

Despite the environmental-friendly characteristic of GPC, good mechanical properties, and high durability, there are several challenges that need to be tackled before it can become a potential mainstream construction material. One such challenge is the need for a high temperature curing regime as reported in literature [6]. However, the main challenge is the dependency of the final product on the activation system and source of the pre-cursor material. There are different types of pre-cursor materials with low and high calcium content [3]. It has been stated in literature that low calcium fly ash has similar properties to OPC concrete, however it requires heat curing for strength development. This research was triggered to take part in overcoming the above-mentioned challenges.

There are several studies available in the literature focusing on the behaviour of heat-cured GPC [10, 13], however, study on the properties of ambient cured GPC is limited [14, 15]. The work presented here is part of an ongoing investigation on the mechanical properties of ambient cured GPC. The key parameters of the study were the binder material content (percentage of FA and GGBFS) and the ratio of sodium silicate to sodium hydroxide (Na_2Sio_3/NaOH).

2. Materials Properties

In-organic by product industrial waste like FA and GGBFS were used as binders in this study. An XRF test was conducted to specify the composition of the pre-cursor materials and the results are reported in Table 1. The alkaline activators used in the mixes were NaOH & Na_2Sio_3 solutions. Na_2SiO_3 solution had 27% of SiO, 8% of Na_2O, and a density between 1.296 and 1.396 gr/ml. NaOH solution (12M) were prepared from 98% purity flakes dissolved in water. The aggregates were combination of coarse and fine. Natural crushed aggregates with maximum size of 10 mm, absorption of 0.7%, and specific gravity of 2.68 was used as coarse aggregates. While fines used in the mix was a 60% - 40% combination of 5mm crushed aggregates and dune sand (absorption of 1% specific gravity of 2.62).

Table 1. XRF (x-ray fluorescence) test results.

Material	Na_2O	MgO	Al_2O_3	SiO_2	SO_3	K_2O	CaO	TiO_2	MnO_2	Fe_2O_3
FA	1.045	7.51	10.97	59.83	1.545	1.42	8.75	1.08	0.175	7.525
GGBFS	-	6.58	5.71	32.36	2.38	-	51.01	0.64	0.73	0.59

3. Specimens Preparation

The mixing procedure of GPC mixes was done at room temperature. First coarse and fine aggregates were mixed with some portion of water then fly ash and GGBFS were added and mixed for two minutes. After that sodium silicate solution was poured to the dry mix and mixed for two minutes. Sodium hydroxide solution was then added to the mix and blended for another two minutes. The remaining of water together with superplasticizer was then added at the end and mixed for additional two minutes. It should be noted that sodium hydroxide solution was prepared at least 24 hrs before casting.

The mix proportions based on the work of Junaid et al [16] were used to cast specimens of this study and are presented in Table 2. A total of three sets containing 15 different mixes of GPC were prepared and cast in 75 cylinders of 100x200 mm. The difference between each set is the Na_2Sio_3/ NaOH ratio (R). Set1 has R=1, Set2 with R=2, and in Set3 R=2.5. In addition, Set3 has less percentage of GGBFS compared to Set1 and Set2. The difference between the mixes within the same set was the percentage of GGBFS added to the mix. For example, Mix1 of Set1 and Set2 has no GGBFS while Mix5 of the corresponding sets has 100% GGBFS but no FA. As mentioned earlier Set3 has less GGBFS content with the maximum of 20 % in Mix5. For details of other mixes refer to Table 2. In Table 2 each mix is identified by a specific label representing the designated number of that mix and the number of corresponding sets. For example, S3M1 stands for the GPC Mix1 related to Set3. The water to binder ratio (w/b) was constant at 0.32 for all mixes of the three sets. To calculate the water to binder ratio the following equation was used:

$$w/b = \frac{W_{free} + W_{OH} + W_{Si}}{FA + G + S_{OH} + S_{Si}} \tag{1}$$

Where, W_{free} is mass of free water in the mix, W_{OH} is the hydroxide solution water content, W_{si} is the water content of Na_2SiO_3 solution, *FA* is the mass of fly ash, *G* is the mass of GGBFS, S_{OH} is the mass of solid content of hydroxide solution, and S_{si} is mass of solid content in the Na_2SiO_3 solution

The cylinders were filled with GPC in three layers and was compacted by manual strokes with a 20 mm steel rod and also vibration table was used for better compaction as shown in Figure 1. The samples were demoulded after 24 hours of casting and placed inside laboratory with a room temperature and RH of 40% until it was tested. It is worth mentioning that during casting the setting times of mixes that had higher percentage of GGBFS was less than the mixes with lower amount of GGBFS.

Table 2. Mix proportions of the GPC mixes studied in this paper.

Mix ID		Pre-cursor Materials		Activators		Water
		FA	GGBFS	NaOH	Na_2Sio_3	Potable
	S1M1	425	0	82.5	82.5	58.5
	S1M2	340	85	82.5	82.5	58.5
Set1	S1M3	212.5	212.5	82.5	82.5	58.5
	S1M4	85	340	82.5	82.5	58.5
	S1M5	0	425	82.5	82.5	58.5
	S2M1	425	0	55	110	60
	S2M2	340	85	55	110	60
Set2	S2M3	212.5	212.5	55	110	60
	S2M4	85	340	55	110	60
	S2M5	0	425	55	110	60
	S3M1	425	0	47	118	60
	S3M2	403.7	21.3	47	118	60
Set3	S3M3	382.5	42.5	47	118	60
	S3M4	361.3	63.7	47	118	60
	S3M5	340	85	47	118	60

Aggregates were added with the same proportion in all the mixes (10mm size = 1065 kg/m^3, 5mm size = 390 kg/m^3, and dune sand = 260 kg/m^3

(a) (b)

Figure 1. Samples preparation; a) casting GPC cylinders and b) compaction of GPC cylinders.

IOP Conf. Series: Materials Science and Engineering **839** (2020) 012005 doi:10.1088/1757-899X/839/1/012005

4. Results and Discussion

4.1. Compressive Strength

Compression tests were conducted as per the ASTM C39/C39M-14 standard. The samples were loaded with a pace rate of 2.4 kN/s until failure. Figure 2 shows the average seven and 28-days compressive strength of three samples from each mix.

As can be seen in Figure 2a, for Set1 (with R = Na_2Sio3 / NaOH = 1) with an increase in percentage of GGBFS the compressive strength also increased and reached a maximum value of around 45 MPa at 28 days for mix with 80 % of GGBFS. However, there is no noticeable increase in compressive strength of Mix5 compared to Mix4. This indicates that the strength development is not significant by increasing the percentage of GGBFS beyond 80%. On the other hand, the compressive strength of the samples for mixes without GGBFS was less than 5 MPa in all Sets indicating that the mixes with only fly ash is not appropriate for structural concrete.

The average seven- and 28-days compressive strength of Set2 (with R = Na_2Sio_3 / NaOH = 2) followed a similar trend as Set1 (see Figure 2b). The samples in this set exhibited higher compressive strength value compared to the samples in Set1 due to the higher amount of silicate solution. Therefore, Mix4 and 5 of Set2 returned a maximum strength of 55 MPa at 28 days which is 10 MPa higher than the corresponding mixes of Set 1. The rate of strength gain with respect to the percentage of GGBFS for this set was a bit steeper than Set1. Nonetheless, the compressive strength of S2M1 and S2M2 are similar to S1M1 and S1M2 and were the lowest in these sets. It may, therefore, be suggested that the mixes with very low proportions of GGBFS are not appropriate for applications in structural engineering.

Mixes in Set3 (with R = Na_2SiO_3 / NaOH = 2.5) had lower GGBFS percentage (0, 5, 10, 15 and 20 % respectively) compared to the first two Sets (Set1 and Set2 (0, 20, 50, 80, 100%)). As plotted in Figure 2c, the pattern of strength improvement with respect to the percentage of GGBFS available in the mix was almost similar to Set1 and Set2. However the measured compressive strength is quite low which is not suitable for structural applications. The reason for having overall lower strength in this set is the amount of GGBFS content available in the mixes. Even though there is a relatively steep jump in the compressive strength of Mix4 and 5 of Set3 but still not a suitabel compressive strength. The maximum GGBFS content in this Set was 20% (S3M5) which is equal to the amount of GGBFS in S1M2 and S2M2. This indicates that for camparion purpose Mix5 of Set3 can be compared to Mix2 of Set1 and Set2. When comparing it can be found that S3M5 exhibited higher compressive strength than S1M2 and S2M2 indicating the effect of higher amount of silicate solution (R = 2.5) on compressive strength. It must be noted that higher percentages of GGBFS in mixes with R = 2.5 resulted in early and sudden setting of the GPC mixes.

In general, it can be inferred from Figure 2 that the average 28-day compressive strength of samples of all sets follow a similar trend to the strength acquired at seven days. However, there is not much gain in strength when curing is increased from seven to 28 days. Therefore, the difference between the seven- and 28-days compressive strength of all the mixes is minor. This agrees with the finding in literature. Earlier research work [17], suggest that strength of GPC develop in the initial days after casting and no further improvement is expected at later stages. In addition, the results of this study may also suggest that strength gain with time depend on the type of precursor materials and type of activator used. For example, from Mix3 and onwards that has higher GGBFS content, the difference in strength of seven and 28 days is a bit higher than the first two mixes of each set, however, to reach to concrete conclusion further studies need to be conducted.

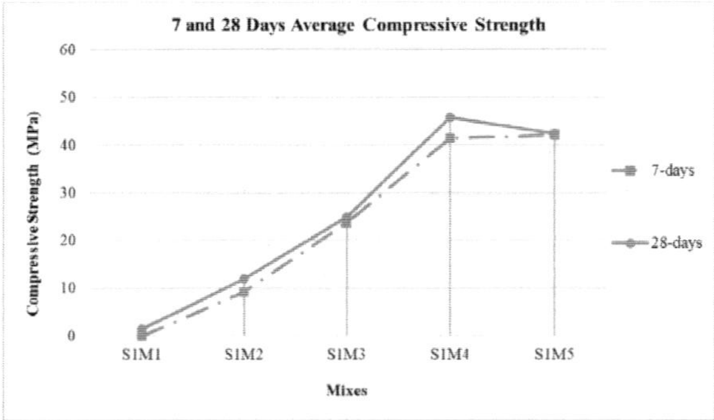

(a) Set1

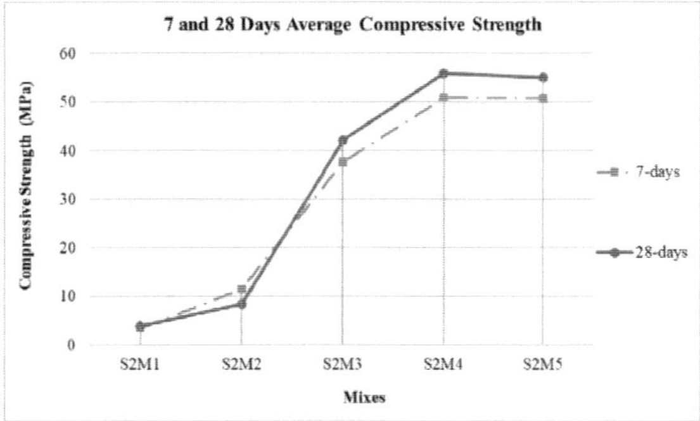

(b) Set2

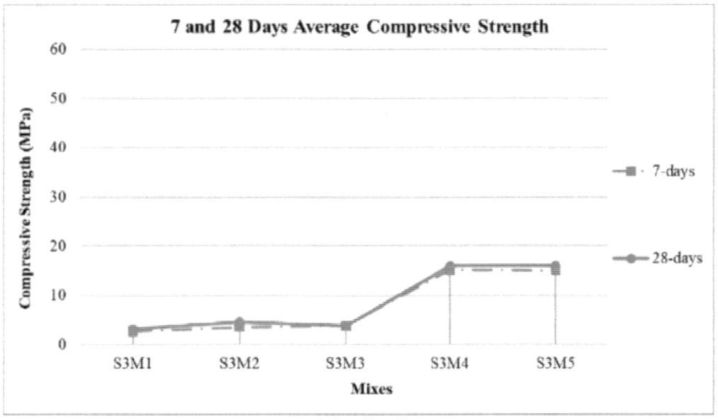

(c) Set3

Figure 2. Compressive Strength of the Samples at the age of 7 and 28 days, a) Set1, b) Set2 and c) Set3.

4.2. Stress – Strain Curves

Along with the compressive strength test, the stress-strain behaviour of selected mixes was also studied. From each set the mixes with strength over 15MPa were selected for this purpose. The results of the tests are plotted in Figure 3 a-c. Each curve is representative of two cylinders tested at the age of 28 days in a displacement-controlled loading mode with a loading rate of 0.01mm/sec until reaching half the ultimate capacity of the sample (Figure 4). After which, the loading rate was reduced to 0.005 mm/sec till failure of the sample. The parameters that can be determined from the stress-strain plots are the maximum stress, strain at maximum stress, the maximum strain, and sometimes the residual stress.

As can be seen in Figure 3, the stress-strain behaviour of the mixes within the same set is different due to variation in their compressive strength. The two mixes labelled S1M4 and S1M5, shows a strain magnitude of 0.003 at the maximum load similar to OPC concrete (see Figure 3a). Similarly, the mixes in Set2 that had higher compressive strength (S2M3, S2M4, and S2M5) showed a strain magnitude of around 0.004 at the time of maximum load as shown in Figure 3b which is slightly higher than the strain in OPC concrete. Mixes in Set3 that has overall lower compressive strength, experienced higher level of strains compared to Set1 and Set2 as expected (see Figure 3c). Furthermore, it can be noticed in Figures 3a and 3b that a higher level of FA replacement with GGBFS gives a higher compressive stress but a lower peak strain.

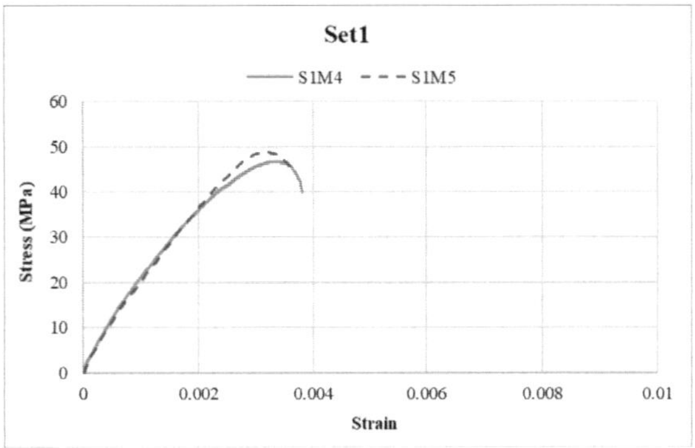

(a)

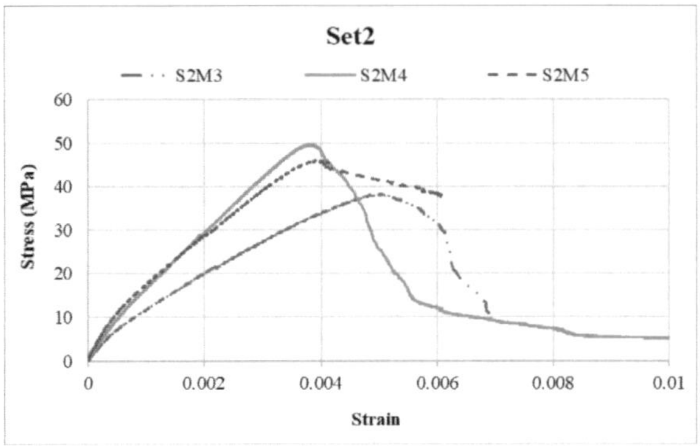

(b)

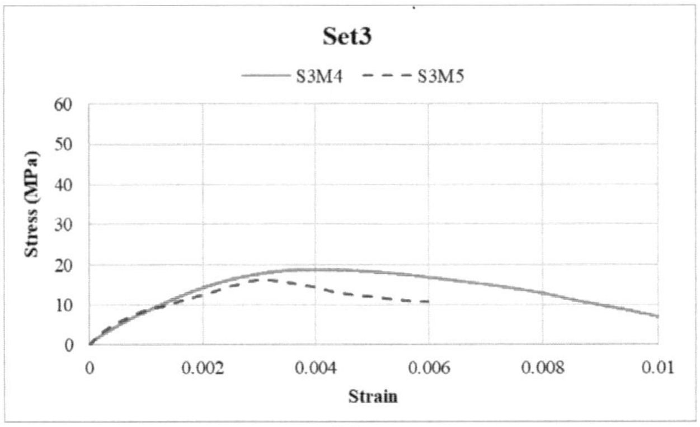

(c)

Figure 3. Stress-Strain behaviour of samples, a) Set1, b) Set2 and c) Set3.

(a)

(b)

Figure 4. GPC cylinder under test; a) compression test b) stress-strain test.

To compare the stress strain behaviour of the mixes investigated in this study with model prediction, two analytical models were considered namely Collins & Mitchel model [18] and Junaid et al. model [12]. Both models take the following form:

$$\sigma_c = f'c \cdot \frac{\varepsilon_c}{\varepsilon_{cm}} \cdot \frac{n}{n - 1 + ({}^{\varepsilon_c}/_{\varepsilon_{cm}})^{nk}} \tag{2}$$

Where $\mathcal{E}_c$ is strain at stress σ_c, $f'c$ is peak compressive strength, $\mathcal{E}_{cm}$ strain at maximum strength. In the Collin model the factors n and k can be found using the following expressions:

$$n = 0.8 + (f'c/17) \tag{3}$$

$$k = 0.67 + (f'c/62) \text{ if } \mathcal{E}_c / \mathcal{E}_{cm} > 1,\ k = 1 \text{ if } \mathcal{E}_c / \mathcal{E}_{cm} \leq 1 \tag{4}$$

While the Junaid et al. model proposes the following values for n and k factors:

$$n = 0.7 + (f_c'/23) \quad k = 0.6 + (f_c'/86) \text{ when } \varepsilon_c / \varepsilon_{cm} > 1 \quad k = 1.0 \text{ when } \varepsilon_c / \varepsilon_{cm} \leq 1 \tag{5}$$

In this study a post peak reduction factor of 2.0 was used for Junaid et al model.

Figure 5 shows experimental data obtained by testing GPC cylinders and the prediction by Collins & Mitchel model and Junaid et al. model. From each set only one mix is considered for the comparison with the models' prediction. It can be inferred from the figure that both models' prediction is reasonable in the ascending branch. Looking at the descending part, Juniad et al. [12] model seems more accurate.

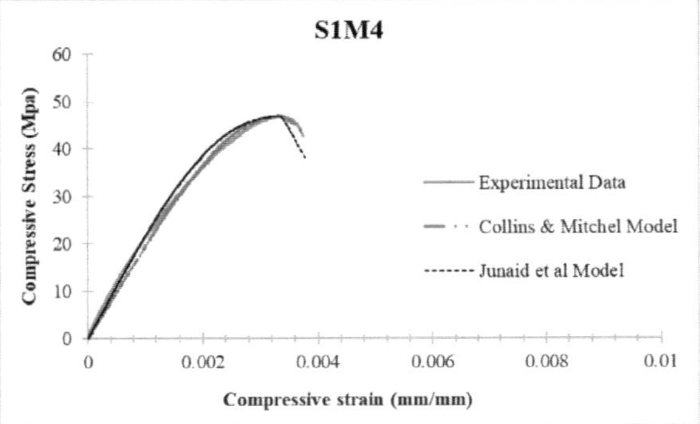

(a)

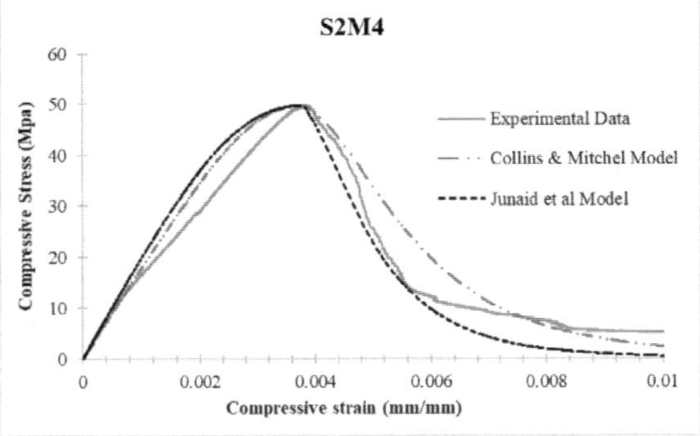

(b)

Figure 5. Comparison of experimental result of Stress-Strain behaviour to the models prediction; a) mix S1M4 and b) mix S2M4.

5. Summary and Conclusions

A total of three sets comprise of 15 different GPC mixes were studied in this research project. The objective was to investigate the strength characteristics of ambient cured GPC and their dependency on activator ratios, the precursor type & proportion, and curing time. The compressive strength along with stress-strain relationship for these mixes was studied. The results found from this investigation shows the feasibility of GPC production. The results also indicate that ambient cured GPC can be made with strength properties similar to ordinary Portland cement concrete.

Based on the results of this study the compressive strength of ambient cured Geo-polymer concrete is highly influenced by both the type and ratios of the precursor material. Having higher percentage of GGBFS in the mix significantly increased the compressive strength but at the loss of setting time. The mixes that had GGBFS around 20% of the binder material returned a compressive strength of less than 25 MPa which may not be suitable for structural applications. Whereas, the increase in percentage of GGBFS from 20 to 80% resulted in compressive strength of 40 MPa and higher. However further increase beyond 80% GGBFS had no impact on the compressive strength. Furthermore, the ratio of activator solutions has a noticeable effect on the strength properties. In all three sets with different R = Na_2Sio3 / NaOH, among the three mixes that had equal amount of GGBFS (20 %) as binder, the mix that had higher R ratio, exhibited a higher compressive strength. The stress-strain relationship of GPC can be predicted by several analytical models. In this study two such models were used to predict the stress strain behaviour with satisfactory level of accuracy.

6. References

[1] Malhotra V M and Mehta P K 2012 *High-Performance High-Volume FlyAsh Concrete.4th edition*, https://tarjomefa.com/wp-content/uploads/2016/08/5106-English.pdf.
[2] Topark-Ngarm P, Chindaprasirt P and Sata V 2015 Setting Time, Strength, and Bond of High-Calcium Fly Ash Geopolymer Concrete *J. Mater. Civ. Eng.* **27**
[3] Yadollahi M M and Benli A 2017 Stress-strain behavior of geopolymer under uniaxial compression *Comput. Concr.*, **20** 381–89
[4] Davidovits J 1994 Global Warming Impact on the Cement and Aggregate Industries *World Resource Review* **6** 263-78
[5] E. Worrell, L. Price, N. Martin, C. Hendriks and L.O. Meida 2001 Carbon Dioxide Emissions from the Global Cement Industry *Annu. Rev. Energy Environ.* **26** 303–29,
[6] Shehab H K Eisa A S and Wahba A M 2016 Mechanical properties of fly ash based geopolymer concrete with full and partial cement replacement *Constr. Build. Mater.* **126** 560–65
[7] Kristiawan S. A 2018 Uniaxial Compressive Stress-Strain Behavior of Self-Compacting Concrete With High-Volume Fly Ash *Int. J. Geomate* **14** 77–85
[8] Davidovits J 1991 Geopolymers - Inorganic polymeric new materials *J. Therm. Anal.* **37** 1633–1656
[9] McCaffrey R 2002 Climate change and the cement industry *GCL Mag.* https://thegreencaiman.files.wordpress.com/2016/03/climate_change_and_the_cement_industry.pdf
[10] Fernández-jiménez A 2006 Engineering Properties of Alkali-Activated Fly Ash Concrete *ACI Mater. J.* **103**
[11] Lloyd N A and V Rangan B 2010 Geopolymer concrete with fly ash *Sustain. Constr. Mater. Technol.* **7** 1–12
[12] Junaid T M Kayali O and Khennane A 2017 Response of alkali activated low calcium fly-ash based geopolymer concrete under compressive load at elevated temperatures *Mater. Struct. Constr.* **50** 1–10
[13] M.T. Junaid, A. Khennane, O. Kayali, A. Sadaoui, D. Picard and M. Fafard 2014 Aspects of the deformational behaviour of alkali activated fly ash concrete at elevated temperatures *Cem. Concr. Res.* **60** 24–29
[14] Nath P and Sarker P K 2017 Flexural strength and elastic modulus of ambient-cured blended low-calcium fly ash geopolymer concrete *Constr. Build. Mater.* **130** 22–31
[15] Lee N and Lee H 2013 Setting and mechanical properties of alkali-activated fly ash/slag concrete manufactured at room temperature *Constr. Build. Mater.* **47** 1201–09
[16] Junaid M T Kayali O Khennane A and Black J 2015 A mix design procedure for low calcium alkali activated fly ash-based concrete *Constr. Build. Mater.* **79** 301–10
[17] Hardjito D and Rangan B. V 2005 Development and Properties of Low-Calcium Fly Ash-Based Geopolymer Concrete, Research Report GC 1 Faculty of Engineering Curtin University of Technology Perth, Australia
[18] Collins M and Mitchel D 1993 Structural design considerations for high-strength concrete *Concr. Int.* **15** 27–34

IOP Conf. Series: Materials Science and Engineering **839** (2020) 012006 doi:10.1088/1757-899X/839/1/012006

Photocatalytic activity of transition metals (Mn, Fe, Ag and Ni doped ZnO) nanomaterials synthesised via sol-gel method: Active sites over band gap

A K A Darman[1,3], M L Ibrahim[1,3], H A Rafaie[2], M S Mastuli[1,3] and M F Kasim*[1,3]

[1] Centre for Nanomaterials Research, Insitute of Science, Level 3 Block C, Universiti Teknologi MARA, 40450 Shah Alam, Selangor, Malaysia
[2] School of Physics and Materials studies, Faculty of Applied Sciences, Universiti Teknologi MARA, 40450 Shah Alam, Selangor, Malaysia
[3] School of Chemistry and Environment, Faculty of Applied Sciences, Universiti Teknologi MARA, 40450 Shah Alam, Selangor, Malaysia
muhdfir@uitm.edu.my

Abstract. Dopants with varied stoichiometry values were examined to determine their performances on photocatalysis. The sol-gel method was successfully synthesised using transition metals, namely manganese (Mn), Iron (Fe), silver (Ag) and nickel (Ni), as dopants in ZnO nanostructure. $Zn_{(1-x)}Mn_{(x)}O$, $Zn_{(1-x)}Fe_{(x)}O$, $Zn_{(1-x)}Ag_{(x)}O$ and $Zn_{(1-x)}Ni_{(x)}O$ (x=10%) were characterised by using X-ray diffraction (XRD), BET surface area analysis, temperature-programmed desorption of carbon dioxide (TPD–CO_2) and UV-Vis spectrophotometer. A phase study on Mn- and Fe-doped ZnO nanoparticles confirmed that both the dopants experienced exceptional crystallinity with single-phase but Ag- and Ni- produced a multi-phases structures. Active sites of catalyst is proven to be an additional factor contributing towards photocatalysis in this work besides band gap and surface area. Temperature-programmed desorption of carbon dioxide (TPD–CO_2) showed that the incorporation of Ag into ZnO lattice has enhanced the number of active sites on the surface of the catalyst whereas incorporation of Mn, Fe and Ni in ZnO has reduced it as compared to pure ZnO. Photocatalytic degradation of methyl orange for all stoichiometry of Ag-doped ZnO nanoparticles resulted in lower degradation time. Interestingly, 10% of Ag-doped ZnO achieved the highest degradation rate as much as 97.47 % at 160 minutes, hence emerging as the best amongst all other compositions.

1. Introduction

These photocatalysis is the mechanism by which light irradiation is used to enable the catalyst for chemical reactions to be initiated. This process helps to remove organic compounds on a daily basis and is helpful to the treatment system of wastewater. Industrial effluent produces colorants such as textiles, ceramics, paints, and pulp & paper industries that increase the level of toxicity in water streams. Extensive photocatalytic research studies have covered a number of topics, including the development of photocatalysts, ways to improve the performance of photodegradation, critical variables associated with photocatalytic efficiency, as well as advancement and photocorrosion [1]. The broad range of photocatalysis offers a general solution to the pressing problem of water pollution and offers a contrabalance to environmental safety [2]. The picture has been extensively used for its image degradation function, including zinc oxide (ZnO), titanium dioxide (TiO2), and iron (II) oxide. High photocatalytic activity is achieved by oxide based materials, thus enhancing photocatalytic effects [3]. The association with band gap energy is another problem of photocatalytic action. The importance of band gaps generally has an important role to influence semiconductor output. In order to produce photoexcitement, ZnO has a large and direct band gap, which requires a huge amount of energy of the beam (photons). The recombination phenomenon which results in dissipation of absorbed energy in the shape of lattice vibration (heat) and photon generation (light) means that some carriers cannot participates in photocatalytic activities. ZnO performance as a photocatalyst is degraded by the strong recombination of photogenerated electrons and holes pair. There is thus a need to emphasize the need to incorporate dopants as electron scavengers to overcome this obstacle in the efficient transport and separation of charge carriers in photocatalyst [4]. Doping contributes specifically, by increasing the band gap by three, (i) encouraging adsorption, (ii) enhancing ZnO conductance and carrier flexibility, and (iii) altering ZnO's steering and valence band position. Many ZnO semiconductor components have attracted more attention among many scientists, due to the increased amount of charge trap, which decreases the volume recombination, and also more efficiency in the separation of photogenerated electrons and holes [4]–[6].

It has been stated that the transition metal dopants (Mn, Fe, Ag, Ni) work well for improving photocatalytic efficiency and creating a synergistic effect [7]. Nevertheless, no important study was conducted on changes in the band gap in light of photocatalytic activity. In addition, Pant et al.'s work on band-gap power, expressing Ag-doped ZnO improved photocatalytic activity successfully, is not in detail [5]. Sonal et al. revealed that Ni in ZnO/carbon-nanospheres were better photocatalyzed and yet no specific work on the analysis of band-gap [6]. Meanwhile for Mn/ZnO, successful work has been achieved by Achouri et al. in decreasing the band gap value from 3.06 eV to 2.83 eV but doesn't correlate the work with photocatalytic activity of Mn [8]. Similarly to Fe/ZnO, Khayatian et al. showed there is an increment when doping with Fe, yet no relationship study for band gap and photocatalysis [9]. This research therefore ties a valid theory with photocatalysis-related circumstances of band gap.

Different techniques such as sol-gel [10]–[12], hydrothermal [13]-[15], combustion [16]–[18] and chemical vapor deposition [19], [20] have been used for the preparation and processing of ZnO-doped synthesis. The route of synthesis is important because photocatalytic output varies from one path to another. This analysis uses undoped ZnO, $Zn_{(x)}Mn_{x)}O$, $Zn_{(x)}Fe_{x)}O$, $Zn_{(x)}Ag_{(x)}O$ and $Zn_{(1-x)}Ni_{(x)}O$ (x= 10 %). To date, several studies have successfully investigated photocatalytic dopants [5]-[9], [21]-[22], but none documented the relationship between band gap and photocatalytic. For example, Ahmad et al. conducted an analysis of the band gap, the particle size of the photocatalyst, photocatalyst surface and photocatalytic operation, showing that the greater the band gap is. This seems to be the enhancing element of photocatalytic action [7]. The purpose of this study in the present paper is therefore to bridge the information gap between fundamental work and work on application which attempts to overcome the band gap that had not been addressed much in previous studies [5]-[9], [21]-[22], In the present work, XRD, BET Surface Analysis, TPD-CO2 and UV-Vis Spectrophotometer characterized the synthesized doped nanocatalysts. The photodegradation investigation was performed under UV light.

2. Experimental

2.1. Materials
Zinc acetate dihydrate was purchased from R&M chemicals with 99.5% purity. Manganese (II) acetate 99%, Ferum (II) acetate 95%, Silver (II) acetate 99% and Nickel (II) acetate 98% were purchased from Aldrich. These starting materials were mixed with absolute ethanol AnapuR.

2.2. Photocatalytic activity
Undoped and 10% of M-/ZnO (M= Mn, Fe, Ag and Ni) nanoparticles were synthesised under mild sol-gel treatment method. Zinc acetate dihydrate and Manganese (II) acetate tetrahydrate/ Ferum (II)

acetate/Silver acetate/Nickel (II) acetate tetrahydrate were dissolved under absolute ethanol and was stirred for 2 hours to gain a homogenous mixture. Base (ammonium hydroxide) was applied to increase the pH value to 9 and was accompanied by a heating process at 100 ° C. Slow drying was done and grey precursors were obtained within 24 hours. STA (Jupiter STA449-F5) was further used to analyze nanoparticle specimens. Both TGA and DSC provided the most appropriate temperature to continue with the annealing process. For 3 hours, the precursors are cured at 400 ° C. Next, structural crystallinity tests were performed following the annealing process using XRD (PANanalytical) X'pert Pro powder diffraction equipment. The band gap test, which describes the properties of light absorption, was performed under reflectance (% R) mode using the Perkin Elmer Lambda 950 UV-Vis-NIR spectrophotometer. BELSORP-mini II Instrument by BEL Japan Inc. has measured the surface area. Under the plot of Brunauer-Emmett-Teller (BET), the different surface areas of undoped, Mn-, Fe-, Ag- and Ni / ZnO are plotted. Measurement of active sites was determined using TPD-CO_2.

2.3. Photocatalytic activity
The photocatalytic activity of M-/ZnO nanoparticles was determined by the determination of the decomposition of methyl orange at constant room temperature at each interval. The catalyst loading was 100 mg of M-/ZnO catalyst, with an initial concentration of 10 ppm in a medium beaker containing 100 mL of methyl orange solution. UV-light radiation was turned on at 352 nm wavelength and 8 W wavelength. The dye solution was obtained at every 40-minute interval. The photocatalytic analysis was performed using UV-Vis spectrophotometer under absorbance, (A), mode. The peak of absorption of methyl orange was estimated at 464 nm. Efficiency of photodegradation (percent) was measured with respect to the total photodegradation at each interval. The constant frequency of photodegradation, k, has been determined for all samples. Before photodegradation, absorption regulated graph was created by excluding UV-light irradiation.

3. Results and discussion

3.1. Phase study
Figure 1 showed the pattern of XRD for M-/ZnO nanostructures varying from 20° to 90°. Excellent crystallinity was achieved as the diffracted peaks showed a good match with the ICDD reference number 01-089-0510 of ZnO wurtzite hexagonal with the P63mc space band. On Mn-and Fe/ZnO, there were no additional diffraction peaks noted as it is well-slotted and completely replaced between the ZnO structures boundaries. Because the ionic radius of Mn^{2+} (0.66 Å) and Fe^{2+} (0.70 Å) is smaller than that of Zn^{2+} (0.74 Å), the dopant atoms integrated in the ZnO lattices did not alter the overall crystal structure.

As for the Ag/ZnO, diffraction peaks of Ag metal are observed in accordance with the standard Ag ICDD reference number 01-087-0717 at (111), (200), (220), and (311) crystal plane. This phenomenon is due to the immense ionic size difference in the ZnO system between Ag^+ ions (1.15 Å) and Zn^{2+} ions. The presence of Ag metal seemed to have an influence on the manufacture of ZnO [23], [24]. While, the diffraction peak found on Ni/ZnO consisting of NiO and Ni metals at (200) and (111), Ni ICDD reference numbers 01-073-1519 and 00-001-1266, respectively. At 10 % Ni doping, phases added has weakened the bonding of Zn-O connections [26].

No shifts are found in the peak positions for Ag and Ni samples based on the XRD trend. The diffraction peaks for both Ag and Ni dopants showed weaker intensity as compared to undoped ZnO which indicates that metal atoms are not well-slotted into the lattice of ZnO, but squeezed in the grain boundaries of ZnO crystallites [25], [26]. Therefore, the Ag- and Ni/ZnO phenomenon was referred to as interstitial
doping.

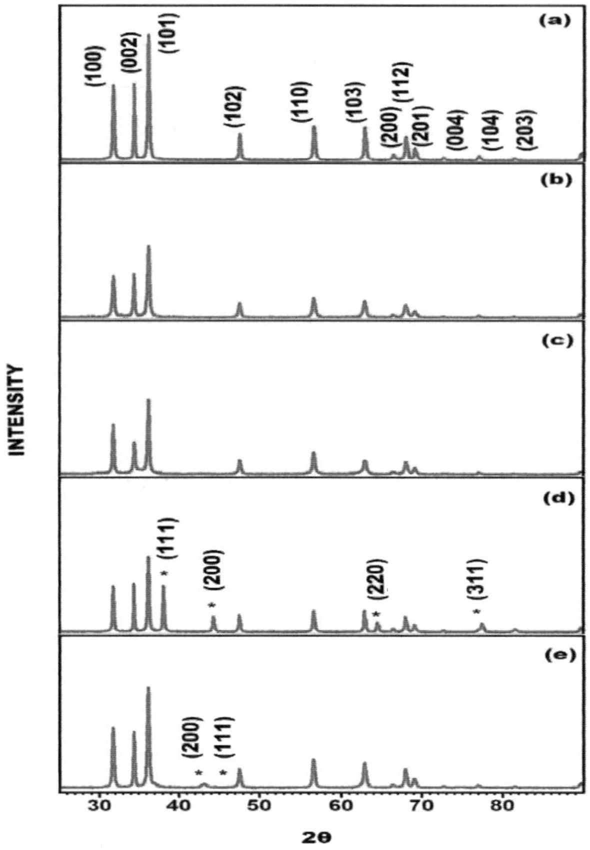

Figure 1. The XRD pattern of **(a)** undoped ZnO **(b)** 10 % Mn/ZnO **(c)** 10 % Fe/ZnO **(d)** 10 % Ag/ZnO and **(e)** 10 % Ni/ZnO materials.

3.2. Band gap study

Figure 2(a) displayed the annealed test optical band gap at 400 °C M-/ZnO. Data absorption edges provide information related to the excitonic and inter-transition of nanomaterials [27], [28]. The change from UV absorption peak spectra observed on the chart from 370 nm to 410 nm is primarily due to the varying sizes of nanomaterials. Reflectance spectra of ZnO is higher than other dopant elements throughout the visible region. Decrease in reflectance spectra may torch better photocatalyst, hence producing better photocatalytic activity theoretically. This decrease is due to various sub-levels through the bandgap of ZnO, giving chances for band gap decrement. There are no fluorescence signal observed for undoped and M-/ZnO after 380 nm indicates the presence of carrier traps within the nanoparticles [10]. Higher particle size of nanocatalyst produces lower surface area, resulting in lower energy value of the band gap, according to Ahmad et al., (2015). Thus, different nanocatalyst sizes can cause the peak to move slightly [22], [28]. The Tauc relationship was also plotted on the basis of the edge of absorption deduced from the spectra of reflection. The Tauc relation was applied via equation below:

$$(\alpha h v) = C(hv - Eg)^n \qquad (1)$$

From the Tauc relationship, α represents the material's absorption coefficient, h denotes Planck's constant, v reflects the light rate, C is the proportionality constant, E_g refers to the band gap power, and n= 1⁄2 (for materials in direct transition mode), as ZnO is known as a direct band gap semiconductor [29]-[31]. In this analysis, the absorption coefficient was calculated by:

$$\alpha = k \ln\left(\frac{R_{max} - R_{min}}{R - R_{min}}\right) \quad (2)$$

k is a constant dependent on the absorption coefficient, R_{max} stands for total reflection, and R_{min} corresponds to minimal reflection. Equation (1) and (2) give the following results:

$$(\alpha h v)^2 = C'\left(hv - E_g\right) \quad (3)$$

Extrapolation was obtained from the $(\alpha hv)^2$ graph against hv, while band gap energy was measured once it reached the x-axis, abscissa line. For M-/ZnO, respectively, the Tauc plot graph and the extrapolation are shown in Figure 2(b). The values of the band gap are shown in Table 1. Change of absorption edge and band gap occurred at interstitial doping is due to metal substitution at the Zn^{2+} site. For example, Ni content in the ZnO system resulted in the presence of the NiO phase due to incomplete substitution. This situation weakened Zn-O bonding, but activated Ni-O growth [26], [32]. The improvements observed in the Ag-and Ni-doped ZnO band gap values are due to grain size, structural factor, and concentration of carrier [33], [34]. Dopants such as Mn, Ag and Ni have larger band gap than undoped ZnO because ions form each representative dopant replaced Zn^{2+} lattice and contributed to free electrons which eventually makes Fermi level going towards or be as same level as conduction band [11].

The band gap has a major role to play in evaluating the efficiency of photocatalysis. This research proves that other factors, as described above, may reduce the band gap as the main factor for photodegradation of methyl orange, as elaborated in the following section.

Table 1. Band gap energy values of undoped and M-/ZnO with different stoichiometry.

Sample	Band gap enery value (eV)
Undoped ZnO	3.29
10 % Mn/ZnO	3.33
10 % Fe/ZnO	3.27
10 % Ag/ZnO	3.31
10 % Ni/ZnO	3.35

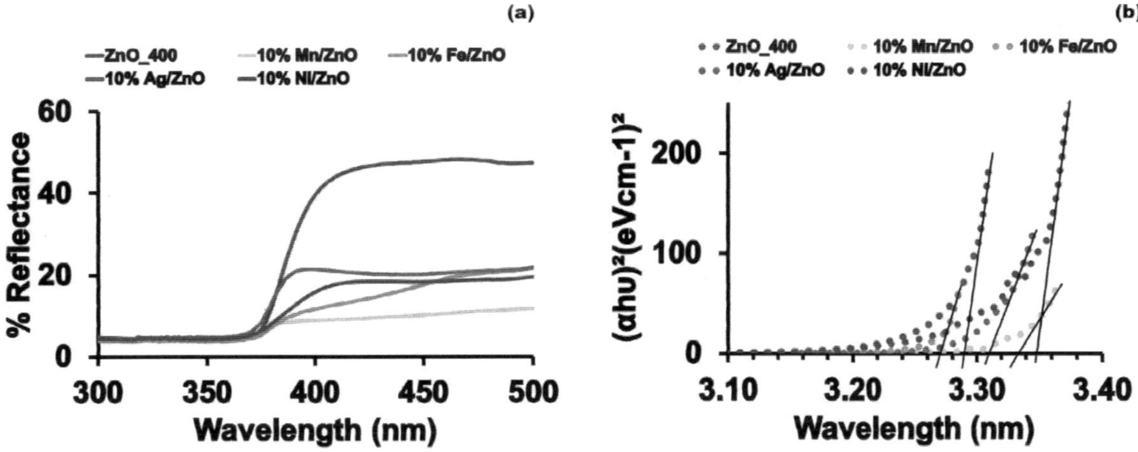

Figure 2. The result of **(a)** UV-Visible spectra and **(b)** Tauc plots of M-/ZnO on each dopant.

3.3. Surface area analysis

Table 2 shows the distribution on undoped and M-/ZnO of N_2 adsorption-desorption and pore length. It is possible to classify all adsorption isotherms as Type II based on the International Union of Pure and Applied Chemistry (IUPAC) classification standard. The porosity reveals its macroporous nature under

Type II conditions. Building Brunauer-Emmett-Teller (BET) plot calculated the approximate surface area, maximum pore density and average pore diameter of undoped and M-/ZnO. Table 2 tabulates the textural properties on each dopant. 10 % Ag/ZnO appeared to result in the smallest specific surface area, and followed by undoped, Mn-, Fe- and Ni/ZnO. As for Ni / ZnO, the smallest size of crystallite was the highest specific area of the layer. High surface area is expected to result in strong photocatalytic behavior in principle. Nonetheless, compared to either undoped or Ag/ZnO, Mn-, Fe-and Ni/ZnO did not generate better photocatalytic activity. In this case, the Mn-, Fe-and Ni / ZnO sites are thought to be inactive against photocatalytic reaction [35], [36].

Table 2. Textural properties of undoped and M-/ZnO.

Sample	$a_{s,BET}$ (m^2g^{-1})	Total pore volume (cm^3g^{-1})	Average pore diameter (nm)
Undoped ZnO	11.671	0.1618	55.448
10 % Mn/ZnO	13.562	0.1257	37.063
10 % Fe/ZnO	13.621	0.0835	24.517
10 % Ag/ZnO	9.7801	0.1363	55.736
10 % Ni/ZnO	14.207	0.1958	56.754

3.4. Active sites measurement

TPD profiles represented the interaction between CO_2 molecules and Table 3 tabulated photocatalyst surfaces. CO_2 desorption occurred at temperatures ranging from 300 °C to 550 °C, demonstrating that all specimens were high basic centers. At temperatures above 350 °C, both undoped ZnO and Ag/ZnO showed a strong desorption maximum. 10 % Ag/ZnO had the highest total volume of basicity (see Figure 3), in which the photocatalyst surface had the highest active sites. Ag/ZnO is therefore a likely place to initiate a good photocatalytic performance compared to Mn-, Fe-and Ni/ZnO. In Mn-, Fe-and Ni/ZnO, asymmetrical, low and decreasing desorption peak was noted due to diffusion limitation and decomposition of species of carbonates. Limitation in diffusion caused by limited distribution of pores on pore volume and dopant Mn, Fe and Ni diameter resulted in low total active site volumes. Although the surface area of Mn-, Fe-and Ni/ZnO (see Table 2) has a slight resemblance to surface area analysis, the sites in them are thought to be less active. Therefore, the photocatalyst surface coverage of reactive adsorbed species is smaller, leading to inadequate photocatalytic efficiency [37]-[39].

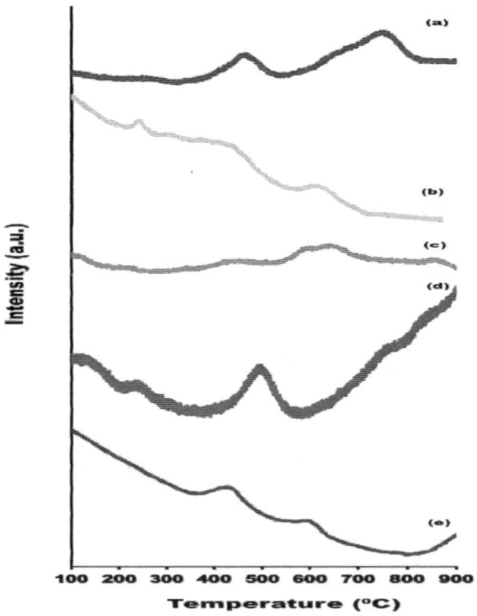

Figure 3. CO_2 – TPD profiles of **(a)** undoped ZnO **(b)** 10 % Mn/ZnO **(c)** 10 % Fe/ZnO **(d)** 10 % Ag/ZnO and **(e)** 10 % Ni/ZnO nanomaterials.

Table 3. Basic sites of undoped and M-/ZnO determined by TPD-CO_2.

Sample	Temperature (°C)	Amount of CO_2 desorbed (μmol/g)	Total amount of basicity (μmol/g)
Undoped ZnO	465	57.92	119.77
	754	61.85	
10 % Mn/ZnO	446	39.39	58.02
	651	18.63	
10 % Fe/ZnO	429	10.74	58.34
	649	47.60	
10 % Ag/ZnO	140	19.35	166.53
	494	147.18	
10 % Ni/ZnO	428	34.08	61.38
	599	27.30	

3.5. Photocatalytic activity

The photodegradation of undoped and M-/ZnO nanomaterials was calculated by evaluating each interval (40 minutes) under 352 nm 8W UV light irradiation of 10 ppm methyl orange aqueous solution. It is revealed, with UV light irradiation, the time-dependent absorbance spectra for undoped and M-/ZnO for 160 minutes from 350 nm to 550 nm. The characteristic value, λmax, was at 464 nm for methyl orange based on the chart. After 160 minutes, undoped and Ag/ZnO showed promising photodegradation with complete clarity, while Mn-, Fe-and Ni/ZnO produced low photocatalytic efficiency.

Photodegradation efficiency (%) was calculated from the graph on the basis of the absorbance measured from each interval. The photocatalytic degradation efficiency is expressed in Equation 4, as follows:

$$\text{Photocatalytic degradation efficiency (\%)} = \left[\frac{(C_0 - C)}{C_0}\right] \times 100 = \left[\left(\frac{A_0 - A}{A_0}\right)\right] \times 100 \qquad (4)$$

C_0 represents the initial concentration of dye from the formula, while C denotes the concentration of dye in terms of time (minutes) at each interval. In the meantime, A_0 is the initial absorption, while A corresponds to the absorption of methyl orange at each interval at a common absorption wavelength of 464 nm. [40], [41].

Figure 4(a) showed the undoped and M-/ZnO photocatalytic behavior that Ag/ZnO tends to be stronger than Mn-, Fe-and Ni / ZnO. 10 % Ag/ZnO resulted in the highest efficiency of degradation with 97.47%. Meanwhile, Mn-, Fe-and Ni/ZnO reported low photocatalytic degradation efficiency. For 10%

of them with 5.48%, 12.31% and 13.44% respectively, no appreciable photodegradation efficiency was observed. The photodegradation rate constant, k, of undoped and M-/ZnO had been assessed using the kinetic model suggested by Langmuir-Hinshelwood, which is pseudo-first kinetics model of photocatalysis [42], as given in Equation 5 below:

$$\ln\left(\frac{C}{C_0}\right) = -kt \tag{5}$$

From the formula, C_0 represents the initial methyl orange concentration, C corresponds to the methyl orange concentration at each interval, and t is the time of irradiation. The ln (C/C_0) graph against t shows a linear relationship in Figure 4(b), followed by kinetics of pseudo-first-order reaction. In Table 4 for undoped and M-/ZnO, the pseudo-first order frequency constant, k, and linear regression, R^2, are tabulated respectively. 10% of Ag content in ZnO gave the highest k values amongst other dopants.

Table 4. Photocatalytic performance of undoped and M-/ZnO nanoparticles.

Sample	Percentage degradation, (%) after 160 minutes	Photodegradation rate constant, (k, min^{-1})	R^2	Band gap energy value (eV)
Undoped ZnO	77.66	0.0088	0.9824	3.29
10 % Mn/ZnO	5.48	0.0004	0.9814	3.33
10 % Fe/ZnO	12.31	0.0007	0.9068	3.27
10 % Ag/ZnO	97.47	0.0233	0.9624	3.31
10 % Ni/ZnO	13.44	0.001	0.9702	3.35

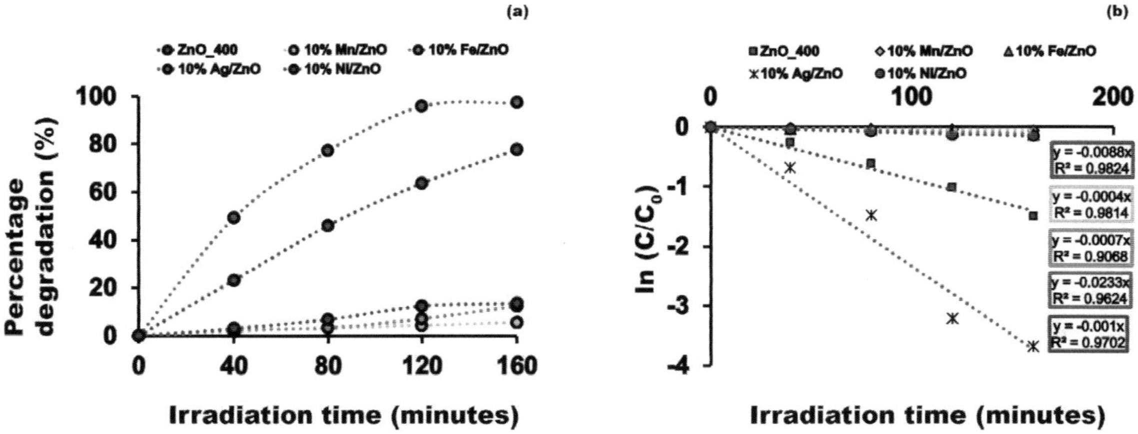

Figure 4. The result of **(a)** photocatalytic degradation efficiency of M-/ZnO and **(b)** curves of ln (C/C_0) vs time.

This research clearly shows that the quality of the M-/ZnO photocatalysis did not significantly affected band gaps. Band gap is the energy gap that represents the existence of photogenerated holes and excited electrons between the valence range and the conduction band. In general, the photocatalytic cycle (see Figure 10) occurs when the energy of a semiconductor, which is ZnO, is more than the origin. Electrons from the valence band (VB) are therefore excited to move to the conduction band (CB).

As an electron scavenger, the dopants (Mn/Fe/Ag/Ni) capture excited electron intercepting recombination of two holes, leading to an extended lifetime of the aroused electron. The reaction to H_2O and exceptional oxidant (•OH, $•O_2^{-}$ and H_2O_2) molecules, which later degrades the dyes, is given greater time for excitable electrons and holes. The oxidant organism has the potential to degrade contaminants by breaking down organic bonds, including the degradation of aromatic rings of dye molecules and their reaction. [3], [43]. Generating more oxidizing organisms improves the efficiency of photocatalytic activity. Ironically, due to the excellent result of the Ag/ZnO nanomaterials, the band gap is not a dominant factor in this work, as it turned out to be a good development in photocatalysis. The most extensive photocatalytic activity among other dopants was 10 % of Ag material. Comparison between 10% Ag-, Mn-, Fe-and Ni/ZnO suggests that band gap is not the factor that controls photocatalytic activity in this work. 10 % Fe/ZnO had a lower energy gap band value compared to 10 % Ag/ZnO, but did not produce better photocatalytic efficiency. This shows that although 10 % Fe/ZnO resulted in the smallest band gap, there was no interconnection with the output of photocatalysis which resulted in low photocatalytic activity. Theoretically, this condition is linked to Ag/ZnO due to the good quality of photodegradation, but not to Fe/ZnO, as well as to Mn-and Ni/ZnO. One of the main reason for the superior photocatalytic quality in 10% Ag/ZnO is that active sites in Ag are available and the suspension has increased light penetration, indicating the plasmonic effect of Ag [44]. In addition, Mn-, Fe-and Ni/ZnO showed photo-phenomena because they showed lower photocatalytic activity because light penetration decreased. Limited access to the catalyst surface via agglomeration affects the adsorption efficiency. Based on the SEM images on Mn-, Fe-and Ni/ZnO, the inability to obtain long nanorods could be the explanation why the light could not penetrate more to its surface. [45], [46].

Dopants (Mn/Fe/Ag/Ni) had almost identical band gap energy values, but vary in terms of output. Therefore, photocatalytic activity cannot be mainly caused by the band gap. Other explanations for influencing the quality of photocatalytic may include surface area [47], [48], particle size [49], crystallinity, and quantum yield [50].

4. Conclusion

Ultimately, the process and performance of photocatalysis concludes fundamental band differences and application work. Synthesis on undoped and 10% M-/ZnO was performed using the modified sol-gel process. The association between band gap and photocatalytic behavior has been shown to be unfounded in this paperwork, as the band gap does not affect photocatalytic efficiency. Other considerations, such as active sites, surface area, quantum yield and crystalline volume, may be more suitable in substituting photocatalysis features. This paper showed that Ag/ZnO only performs better on photocatalysis, but this does not extend to Mn-, Fe-and Ni / ZnO, which equals the band gap value.

5. References

[1] Lee K M, Lai C W, Ngai K S and Juan J C 2016 *Water Res.* **88** 428 - 448
[2] Qi K, Cheng B, Yu J and Ho W 2017 *J. Alloys Compd.* **727** 792 - 820
[3] Rafaie H A, Nor R M, Azmina M S, Ramli N I T and Mohamed R 2017 *J. Environ. Chem. Eng.* **5(4)** 3963–72
[4] Kang X, Liu S, Dai Z, He Y, Song X and Tan Z 2019 *Catalysts* **9(2)** 191
[5] Pant B, Park M, Kim H Y and Park S J 2016 *Synth. Met.* **220** 533–537
[6] Singhal S, Dixit S and Shukla A K 2018 *Adv. Powder Technol.* **29(12)** 3483–92
[7] Türkyılmaz Ş Ş, Güy N and Özacar M 2017 *J. Photoch. Photobio. A: Chem.* **341** 39–50
[8] Achouri F, Corbel S, Balan L, Mozet K, Girot E, Medjahdi G, Said M B, Ghrabi A and Schneider 2016 *Mater. Design* **101** 309–316
[9] Khayatian A, Asgari V, Ramazani A, Akhtarianfar S F, Kashi M A and Safa S 2017 *Mater. Res. Bull.* **94** 77–84
[10] Hasnidawani J N, Azlina H N, Norita H, Bonnia N N, Ratim S and Ali E S 2016 *Procedia Chem.* **19** 211–216
[11] Bekkari R, laânab L, Boyer D, Mahiou R and Jaber B 2017 *Mater. Sci. Semicon. Process.* **71** 181–187
[12] Mahdavi R and Ashraf T S S 2017 *Ultrason. Sonochem.* **39** 504–510
[13] Kumaresan N, Ramamurthi K, Ramesh Babu R, Sethuraman K and Moorthy Babu S 2017 *Appl. Surf. Sci.* **418** 138–146
[14] Ghoderao K P, Jamble S N and Kale R B 2018 *Optik* **156** 758–771

[15] Bazazi S, Arsalani N, Khataee A and Tabrizi A G 2018 *J. Ind. Eng. Chem.* **62** 265–272
[16] Sangeeta M, Karthik K V, Ravishankar R, Anantharaju K S, Nagabhushana H, Jeetendra K, Vidya Y S and Renuka L 2017 Synthesis of ZnO, MgO and ZnO/MgO by solution combustion method: Characterization and Photocatalytic studies *Mater. Today: Proc.*, ed S. Bland (United Kingdom: Elsevier Ltd) vol 4 pp 11791–98
[17] Ajamein H, Haghighi M and Alaei S 2018 *Mater. Res. Bull.* **102** 142–152
[18] Vasei H V, Masoudpanah S M, Adeli M and Aboutalebi M R 2018 *Ceram. Int.* **44(7)** 7741–45
[19] Kumar S, Sahare P D and Kumar S 2018 *Mater. Res. Bull.* **105** 237–245
[20] Narin P, Kutlu E, Atmaca G, Atilgan, A, Yildiz A and Lisesivdin S B 2018 *Optik* **168** 86–91
[21] Siva Vijayakumar T, Karthikeyeni S, Vasanth S, Ganesh A, Bupesh G, Ramesh R, Manim M and Subramanian P 2013 *J. Nanosci.* **2013** 1–7
[22] Ajala F, Hamrouni A, Houas A, Lachheb H, Megna B, Palmisano L and Parrino F 2018 *Appl. Surf. Sci.* **445** 376–382
[23] Thomas M A and Cui J B 2009 *J. Vac. Sci. Technol. B.: Microelectron. Nanometer Struc.* **27(3)** 1673
[24] Wang L N, Zhou L, Zhou D, Yu Y, Luan Z Q, Tang D L, Zhu N, Gao Z H, Liu N and Hu, L Z 2017 Effect of annealing treatment on structure of Ag doped ZnO films deposited by thermal evaporation method *IOP Conf.Mater.Sci.Eng.* (United Kingdom: Institute of Physics) vol 213 p 12024
[25] Mohammad Kakhki R, Tayebee R and Ahsani F 2017 *J. Mater. Sci.: Mater. Electron* **28(8)** 5941–52
[26] Goswami N and Sahai A 2013 *Mater. Res. Bull.* **48(2)** 346–351
[27] Singhal S, Kaur J, Namgyal T and Sharma R 2012 *Physica B: Condens. Matter* **407(8)** 1223–26
[28] Kaneva N V and Dushkin C D 2011 *Bulg. Chem. Commun.* **43(2)** 259–263
[29] Zhang T, Liang R, Dong L, Wang J, Xu J and Pan C 2015 *Nano Res.* **8(8)** 2676–85
[30] Pearton S J, Norton D P, Ip K, Heo Y W and Steiner T 2005 *Prog. Mater. Sci.* **50** 293 - 340
[31] Rusdi R, Rahman A A, Mohamed N S, Kamarudin N and Kamarulzaman N 2011 *Powder Technol.* **210(1)** 18–22
[32] Zhao J, Wang, L, Yan X, Yang Y, Lei Y, Zhou J, Huang Y, Gu Y and Zhang Y 2011 *Mater. Res. Bull.* **46(8)** 1207–10
[33] Veldurthi N K, Jitta R R, Ravi G, Guje R, Velchuri R, Venkataswamy P and Vithal M 2016 *ChemistrySelect* **1(11)** 2783–91
[34] Naveen Kumar S K, Gayithri K C and Kiran S 2015 *Int. J. ChemTech Res.* **7(2)** 936–942
[35] Iwase M, Yamada, K, Kuri T, Ohtani B and Wakita H 2013 *Appl. Catal. B.: Environ.* **140–141** 327–332
[36] Liao C Y, Wang H P and Lin H P 2013 *Int. J. Photoenergy* **2013** 243160
[37] Guo J, Wang K and Wang X 2017 *Catal. Sci. Technol.* **7(24)** 6013–25
[38] Farabi M S A, Ibrahim M L, Rashid U and Taufiq-Yap Y H 2019 *Energy Convers. Manag.* **181** 562–570
[39] Natesakhawat S, Ohodnicki P R, Howard B H, Lekse J W, Baltrus J P and Matranga C 2013 *Top. Catal.***56** 1752–63
[40] Bai Y, Chen S, Lu L and Bao J 2013 *Adv. Mater. Res.* **602–604** 169–173
[41] Aggarwal S 2016 *Int. Res. J. Eng. Technol.* **3** 2–6
[42] Magalhães P, Andrade L, Nunes O C and Mendes A 2017 *Rev. Adv. Mater. Sci.* **51(2)** 91–129
[43] Kumar R, Umar A, Kumar G, Akhtar M S, Wang Y and Kim S H 2015 *Ceram. Int.* **41(6)** 7773–7782
[44] Zhao Z J, Hwang S H, Jeon S, Hwang B, Jung J Y, Lee J, Park S H and Jeong J H 2017 *Sci. Rep.* **7(1)** 8915
[45] Mekasuwandumrong O, Pawinrat P, Praserthdam P and Panpranot J 2010 *Chem. Eng. J.* **164(1)** 77–84

[46] Grzechulska J and Morawski A W 2002 *Appl. Catal. B.: Environ.* **36(1)** 45–51
[47] Kaur J and Singhal S 2014 *Ceram. Int.* **40(5)** 7417–24
[48] Reinosa J J, Leret P, Álvarez-Docio C M, Del Campo A and Fernández J F 2016 *Bol. Soc. Esp. Ceram. V.* **55(2)** 55–62
[49] Han R, Wang Y, Yu W, Zou W, Shi J and Liu H 2007 *J. Hazard. Mater.* **141(3)** 713–718
[50] Lv J, Gong W, Huang K, Zhu J, Meng F, Song X and Sun Z 2011 *Superlattice Microst.* **50(2)** 98–106

Acknowledgments

The authors wish to thank the internal grant agency funded by Universiti Teknologi MARA, Malaysia, under the UiTM Research Grant No. 600-IRMI/DANA KCM 5/3/LESTARI (103/2017). Special gratitude to Institute of Science, UiTM for provision of research instruments.

IOP Conf. Series: Materials Science and Engineering **839** (2020) 012007 doi:10.1088/1757-899X/839/1/012007

Stress corrosion cracking for dissimilar capacitive discharge welding joint with varied surface preparation and inputted energy

Osmar B Lobo[1], Djarot B Darmadi[2] and Oyong Novareza[3]

[1] Mechanical Engineering Departement, State Polytechnic of Fakfak, Indonesia
[2] Mechanical Engineering Department, Brawijaya University, Indonesia
[3] Industrial Engineering Department, Brawijaya University, Indonesia
buntulobo.osmar@gmail.com

Abstract. In this article is presented dissimilar welding joint that joins steel and brass bars. Due to its varied melting point the welding process of dissimilar metal has a higher degree of difficulties. Naturally, welding process embeds residual stress and provides voids at least in micro scale. The voids then act as a trigger to develop initial cracks which will grow under external load. Once, the welding joint is exposed to corrosive environment, it will be susceptible to Stress-Corrosion-Cracking (SCC) failure. The quality of the welded joints is considered based on SCC resilience. Two parameters of the Capacitive Discharge Welding Joint which have prominent effects were evaluated: surface preparation and inputted energy. With a proper surface preparation it is hoped optimum welded joints can be obtained. The experiment results show that the α_1 surface preparation, i.e. the sharpened steel interface while the brass is kept on perpendicular edge, provides better performance under SCC load. Highest inputted energy (120J) always provides the longest duration for joint exposed to SCC load before the failure. The high inputted energy gives high heat to melt the interface which in turn produces compact joint at the interface.

1. Introduction

There are method of joint can be found such as rivet, bolt or welding. Welding may the most prominent and widely used due to its effectiveness: easy to operate, high strength and avoid water leak. There are many welding method have been developed and one of them is Capacitive Discharge Welding (CDW). CDW is electric resistance method in which the high pulse electric current produce by high capacitor bank [1]. Scothmer mentioned automotive some industry in United States is starting to apply the CDW [2] which is endorsed by Panella since CDW provide fast and reproducible process [1]. Chiozzi modified the interface of the welded specimen. The waving interface was implemented to increase the area of fusion and improve the weldability [3]. Dattoma carried out research to obtain optimum welding parameter and surface of CDW. In this research activity, 150 cylindrical specimens with bore diameter 6 mm and different igniter dimensions were machined in AISI 304 austenitic stainless steel. Parameters and geometry igniter / tip were classified into 15 classes. The results shown that the class 14 geometric igniter : height / h = 0.7mm, diameter / d = 2mm, and welding parameters were: compressive force / F = 65kg, energy input / P = 45Kj was the most optimal [4].

The joint may be needed for different materials which for welding it is known as dissimilar welding. The dissimilar welding is applied when different properties are needed [5]. The joint of dissimilar metals indeed is more complicated compared to same metals due to variation in physical, metallurgical and mechanical properties of the joined metals [6].

Ozlati welded carbon steel and austenitic stainless steel which can be considered as dissimilar welding. The obtained result was very good since the tensile test shown that the highest tensile strength was at welding area. The resulted flash also very neat which reduce the additional process after welding joint is obtained [7]. This research also observes the dissimilar welding joint by means Capacitive Discharge Welding.

Stress Corrosion Cracking describes the failure of materials due to the crack growth as a result of synergy of external mechanical load and corrosive environment [8, 9]. SCC caused catastrophic failures at the constructions, air plane and pipeline [10–12]. SCC was considered as the main phenomenon in the pipelines leaks and blow out with disaster consequences [13]. For certain materials the SCC depend on the aggressive corrosion environment and residual stress [14]. Residual stress is always embedded in the welding joint. Voids which are produced in the welding process trigger the development of cracks. Due to the importance to consider the SCC resilience of welding joint, the resulted welding joint will be evaluated from the SCC point of view.

2. Method

The dissimilar metal joint of steel and brass is obtained using CDW. Based on previous literature there are various geometry when stud or pins welding was applied [15–17]. Hou discussed varied tip geometry for dissimilar aluminum friction welding [18]. The interfaces or tips also been altered for varied welding process [1, 4, 19–21]. Varied surface preparation by means of geometry in this research are engineered as shown in Figure 1a up to Figure 1c, namely specimen α_0, α_1 and α_2.

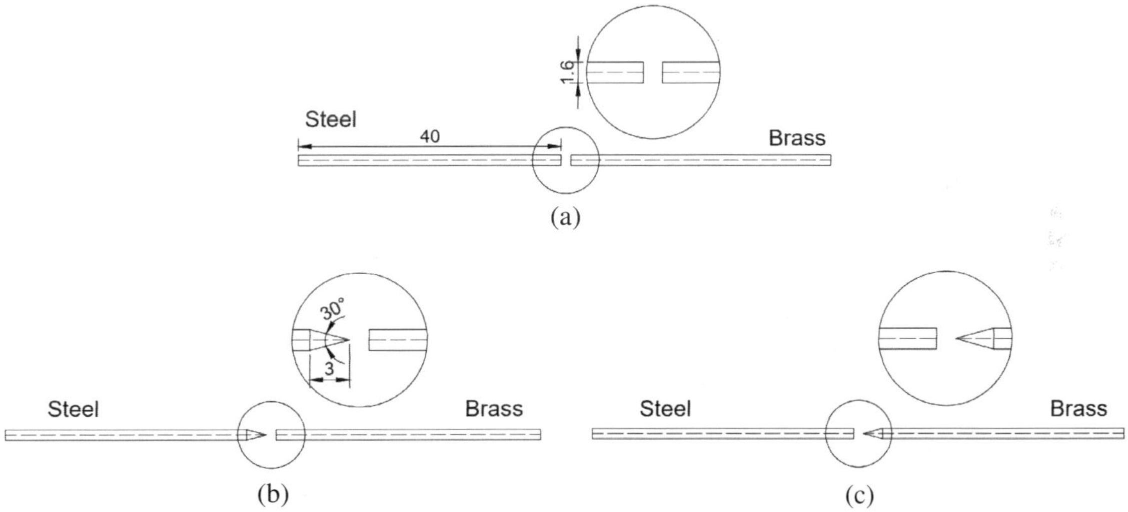

Figure 1. (a) Specimen α_0, (b) Specimen α_1 and, (c) Specimen α_2.

In this research, there was no specimen with sharpened edge on both sides since it is difficult to provide joint with pointed edges on both side due to misalignment. In Figure 2 is shown the schematic diagram and the photo of the CDW process.

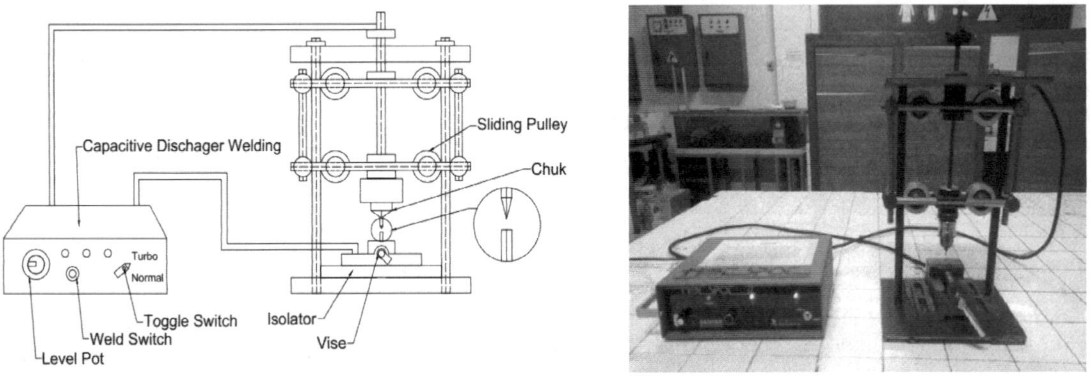

(a) (b)

Figure 2. The CDW Process (a) schematically and, (b) the photo.

The composition of the metals for dissimilar joint is shown in Table 1. In Table 2 is tabulated the mechanical properties of the metals. Welding parameters and constants are listed as in Table 3 considering the manual of the hot-spot CDW.

Table 1. The chemical composition carbon steel and brass rods used in this study.

Materials	Composition (weight %)						
	Fe	C	Si	Mn	Cu	Zn	Sn
Steel	Balance	0,1	1	1,40	-	-	-
Brass	-	-	0.3	-	Balance	5.9	0.8

Table 2. The mechanical properties carbon steel and brass rods used in this study.

Materials	Mechanical Properties			
	Melting point ^{0}C	UTS N/mm^2	Hardness BHN	Density (g/cm^3)
Steel	1450	1036	120	-
Brass	899	888	-	8.4

Table 3. Parameters used for this study.

No	Parameters	
1	Electrode pressure	39,2 N
2	Weld Energy	80, 90, 100, 110, 120 J
3	Spesimen diameter	1.6 mm
4	Voltage	75 VDC
5	Drop height	5 mm

After the dissimilar joint with varied condition is obtained, the Stress-Corrositon-Cracking test was applied. The SCC test was carried out in laboratory scale. The testing apparatus for SCC was design deliberately for this research using Constant Load Test (CLT). In his research, Chida compared varied SCC test those are CLT, Slow Strain Rate Test (SSRT) and Conventional Strain Rate Test (CSRT) which provide equal results [22]. CLT is used in this research due to in simplicity and easy to be made using typical materials. The drawbacks of the CLT is the quite long duration of the experiment. The sketch of the apparatus shown in Figure 3. The constant dead load was applied at one edge which provides tensile load at specimen on another edge.

IOP Conf. Series: Materials Science and Engineering **839** (2020) 012007 doi:10.1088/1757-899X/839/1/012007

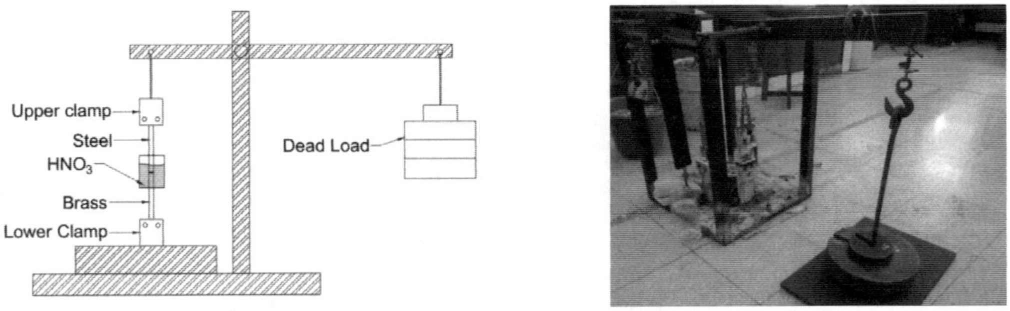

(a) (b)

Figure 3. Constant load test installation (a) schematically and, (b) the photo.

3. Results and Discussion
The main results of this research is the duration of the dissimilar joint exposed to SCC load which can be seen in Table 4.

Table 4. The stress corrosion resilience.

Weld Energy	Time to failure (minutes)		
	α_0	α_1	α2
80J	00:52:40	00:59:49	00:55:52
90J	00:56:04	01:08:18	00:59:13
100J	00:58:17	01:12:51	01:05:46
110J	01:01:03	01:16:26	01:10:03
120J	01:05:53	01:21:45	01:15:19

The independence variables are inputted energy namely 80 J, 90 J, 100 J, 110 J and 120 J; and surface preparation, i.e., α_0 and α_1. The energy is in the range defined by manual book of CDW machine [15]. The physical meanings of the α_0 and α_1 can be confirmed in Figures 1, 2 and 3. The dependence variable are time to failure. The Table 4 is presented in a graphic as shown in Figure 4.

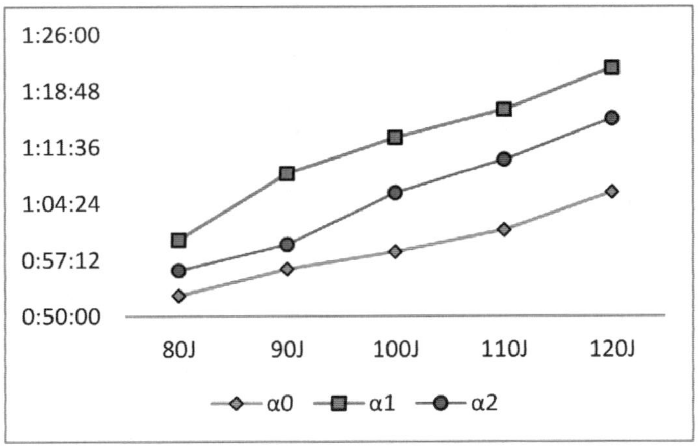

Figure 4. Time to failure for joint with varied surface geometry and inputted energy.

Based on the Figure 5 it can be said that, for certain inputted energy the α_1 specimen always provides better SCC resilience. The α_1 specimen is dissimilar joint with pointed tip for steel whilst brass was kept blunt (perpendicular tip). For the pointed tip of brass and blunt tip of steel produce better joint quality compared to blunt tip for both metals. However, the results of the point tip at brass still below the point tip at steels.

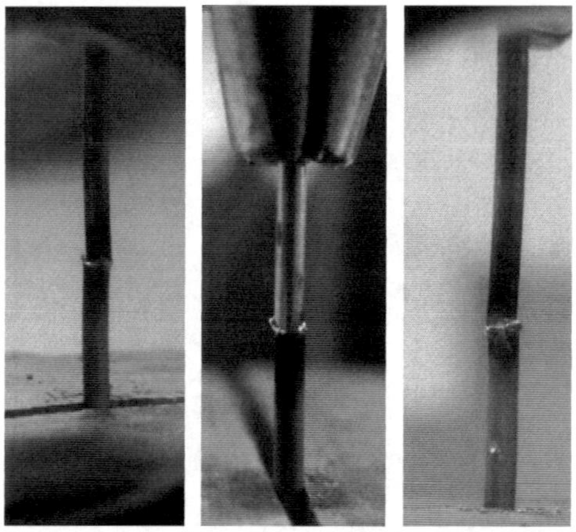

Figure 5. α_0 (left), α_1 (centre) and α_2 (right) for heat input equal to 120 J.

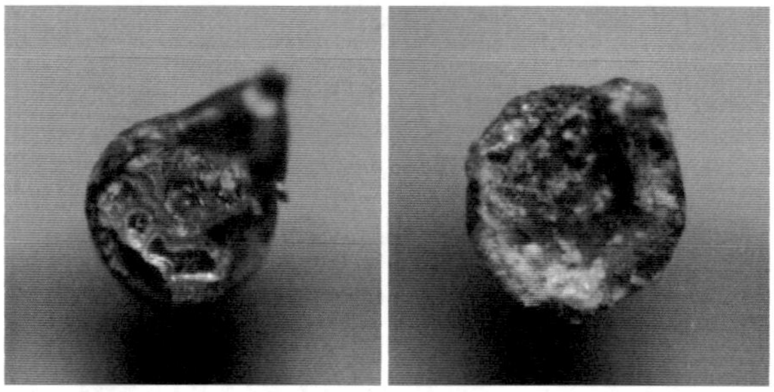

Figure 6. α_0 for heat input equal to 120 J.

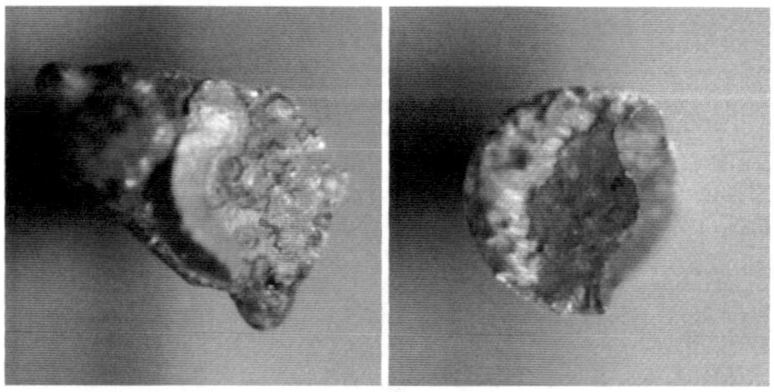

Figure 7. α_1 for heat input equal to 120 J.

Figure 8. α_2 for heat input equal to 120 J.

IOP Conf. Series: Materials Science and Engineering **839** (2020) 012007 doi:10.1088/1757-899X/839/1/012007

Based on macro photos (Figure 6,7 and 8) the flash can be evaluated. Observing the flash it can be predicted the strength of the joint. The thicker flash indicates the strongest joint [23, 24]. From the flash it can be predicted that the same trend also obtained from the developed flash, it is the list of joint for 120J heat input obtained from α_0, α_2, α_1 from the weaker to the stronger joints respectively.

4. Conclusions

Surface preparation does affect the quality of the dissimilar joint by means of SCC resiliencies. The pointed tip of steel provides optimum results. 120 joules is the best welding parameter of all the parameters used in this research. The steel has good weldability with a pointed tip and high weld energy because it will make the steel melted in a higher amount and binds brass more tightly. The lower the energy used, the weld joint will be more susceptible to SCC.

5. Reference

[1] Panella F W, Dattoma V, Giorgi M De, Palano F, Carofalo A 2019 CDW aluminium joints welding and optimisation with NDT/mechanical testing *Int. J. Adv. Manuf. Technol.* 3689–3698

[2] Scotchmer N 2015 The Current Rise in the Use of Capacitor Discharge Welding *Weld. J.* **94** 32–36

[3] Chiozzi S, Dattoma V and Panella F 2008 Capacitor discharge welded bars of Inconel 718 and TiAl6V4 superalloys under fatigue *Mater. Des.* **29** 839–851

[4] Dattoma V, Palano F and Panella F W 2010 Mechanical and technological analysis of AISI 304 butt joints welded with capacitor discharge process *Mater. Des.* **31** 176–184

[5] Harsono D C 2011 Mekanik pada Pengelasan Dissimilar antara Stainless Steel AISI 316L dan Baja Karbon Rendah ST 37 dengan Pengelasan GTAW. Universitas Jember

[6] Abbas E N, Omran S, Alali M, Abass M H and Abood A N 2018 Dissimilar Welding of AISI 309 Stainless Steel to AISI 1020 Carbon Steel Using Arc Stud Welding *Int. Conf. on Advanced Science and Engineering (Duhok, Iraq)* (United States/IEE) p 462–467

[7] Ozlati A, Movahedi M and Mohammadkamal H 2016 Upset Resistance Welding of Carbon Steel to Austenitic Stainless Steel Narrow Rods J. Mater. Eng. Perform. **25** 902–4910 https://doi.org/10.1007/s11665-016-2321-z

[8] Jones R H 1992 *Stress-Corrosion Cracking: Materials Performance and Evaluation* (Ohio: ASM International)

[9] Raja V S and Shoji T 2011 *Stress corrosion cracking-Theory and practice* (New Delhi: Woodhead Publishing Limited)

[10] Darmadi D B, Sugiarto N A and Gapsari F 2018 Stress Corrosion Cracking at ASTM A36 Plate with Varied Grain Orientation *Int. Rev. Mech. Eng.* **12** 987

[11] Sudarsono, Darmadi D B and Gapsari F 2018 Analisis Stress Corrossion Cracking Sambungan Las Gas Tungsten Arc Welding Spesimen Aluminium sebelum dan sesudah Stress Relieving. Universitas Brawijaya

[12] Cottis R A 2000 *Guides to Good Practice in Corrosion Control—Stress Corrosion Cracking* (Teddington, National Physical Laboratory)

[13] National Energy Board 1996 Stress Corrosion Cracking on Canadian Oil and Gas Pipelines.

[14] Cheng Y F 2013 *Stress Corrosion Cracking of Pipelines* (New Jersey: John Wiley & Sons Inc.)

[15] DCC Corporation 2013 Stud Welding Equipment, Stud Welders, Weld Stud Fasteners, Parts, Stud Welder Rentals & Repairs. In: DCC Corp. http://catalog.dcccorporation.com/viewitems/thermocouple-wire/thermocouple-wire-2. Accessed 12 Sep 2019

[16] Sunbelt *SUNBELT Master Catalog-Product Spesification and Aplication Guide* (Texas: Sunbelt Stud Welding, Inc.)

[17] IMAGE The Stud Welding Process. In: IMAGE Ind. INC. https://www.imageindustries.com/what-is-stud-welding/. Accessed 12 Sep 2019

[18] Hou W, Shen Y, Huang G, Yan Y, Guo C and Li J 2018 Dissimilar friction stir welding of aluminum alloys adopting a novel dual-pin tool : Microstructure evolution and mechanical

properties *J. Manuf. Process* **36** 613–620

[19] Casalino G and Panella F W 2006 Numerical simulation of multi-point capacitor discharge welding of AISI 304 bars *J. Eng. Manuf.* **220** 647–655

[20] Casalino G and Panella F W 2007 Microstructural analysis of AISI 304 bars welded with high speed pulsed discharges *J. Mater. Process. Technol.* **191** 149–152

[21] Palano F, Chiozzi S, Panella F W and Dattoma V 2012 Characterization of Steel Welded Joints with Hybrid Projection and Prosesses Mater. Manuf. Prosesses **27** 1387 https://doi.org/10.1080/10426914.2012.663140

[22] Chida T, Hagihara Y, Akiyama E, Iwanaga K and Takagi S 2016 Comparison of Constant Load SSRT and CSRT Methods for Hydrogen Embrittlement Evaluation Using Round Bar Specimens of High Strength Steels *ISIJ Int.* **56** 26–31

[23] Ba¸syi˘git A B and Kurt A 2017 Investigation of the Weld Properties of Dissimilar S32205 Duplex Stainless Steel with AISI 304 Steel Joints Produced by Arc Stud Welding *Metals* **7** 77 https://doi.org/10.3390/met7030077

[24] Connor L P and O'Brien R L 1991 *Welding Handbook: Welding Process* (Miami: Americam Welding Society)

IOP Conf. Series: Materials Science and Engineering **839** (2020) 012008 doi:10.1088/1757-899X/839/1/012008

Wall thickness dependencies of carbon nanotube reinforced nanocomposites

K S Ahmed[1*] , M W Islam[1] and I Ibrahim[1]
[1]Department of Civil Engineering, Military Institute of Science and Technology, Dhaka-1216, Bangladesh
drksa@ce.mist.ac.bd

Abstract. Indentation analyses are conducted to investigate the CNT's wall thickness dependencies of the mechanical properties of nanocomposites. This study incorporates Berkovich nanoindentation analyses for chemically non-bonded CNT/matrix interface, including the size effects of nanocomposites. For non-bonded interface, the properties of nanocomposite are generally controlled by mechanical interlocking, thermal residual stress, Poisson's contraction and van der Waal's (vdW) interaction between CNT and matrix. Considering all these parameters, a series of finite element models for nano-indentation tests have been constructed, aiming to investigate the mechanical properties of CNT reinforced nanocomposites. The model is validated and numerical results are compared with that of existing experimental indentation test. Analysis has been run for ten different wall thicknesses of CNTs in a range of wall thickness starting from 0.034 nm to 0.334 nm. Elasto-plastic behavior of steel and CNT including strain gradient is considered in the program. The study on wall thickness dependencies of CNTs in nanocomposite shows that mechanical properties of nanocomposites largely depends on the wall thickness of CNTs. A minimum wall thickness for a particular tube diameter to achieve a pick value of hardness and modulus of elasticity of nanocomposites is determined. The minimum wall thickness of CNTs is suggested to be 0.2 for an outer diameter of 1nm in order to achieve the maximum value of composite properties. In addition, a wall thickness smaller than 0.05 nm may even reduce the hardness and modulus of elasticity of CNT reinforced composites compare to that of pure matrix.

1. Introduction

Carbon nanotubes have already been considered as one of the super reinforcement materials of high-performance nanocomposites [1-6]. Existing research suggests that the load carrying capability of composites can be enhanced significantly by the addition of carbon nanotubes or nano-ropes [4, 7]. Mechanical properties like modulus of elasticity and hardness of such nanocomposites, are some of the key features that are essential to determine before their application. Numerical indentation test is one of the approaches that is capable of determining mechanical properties of such nanocomposites using indentation model quite accurately [8-11].

Instrumentation for depth sensing indentation procedure was first successfully experimented and developed by Doerner and Nix (1986). The determination of elastic modulus of any solid materials using finite element-based indentation analysis was originally proposed by Larsson et al. (1996). Later, nano-indentation test on polymeric surface was primarily conducted by Briscoe et al. (1998). Based on the indentation technique, Fazio et al. (2001) conducted nano-indentation tests for diamond and hence results were also compared with their finite element model [9].

IOP Conf. Series: Materials Science and Engineering **839** (2020) 012008 doi:10.1088/1757-899X/839/1/012008

This study aims to investigate wall thickness dependencies CNT reinforced composites using FE based nanoindentation simulation considering strain gradient plasticity for chemically non-bonded carbon nanotube reinforced composites. This noble approach would be a breakthrough to simulate indentation test at nanoscale using advanced finite element analysis and hence predict the material properties of nanocomposites for different wall thickness of CNTs before conducting difficult experiments or practical applications.

2. Finite Element Model

In this study, an axisymmetric finite element model has been developed primarily to conduct numerical indentation test as shown in Figure 1. The figure presents a representative FE model of single-wall carbon nanotube (SWCNT) and multi-wall carbon nanotube (MWCNT) reinforced nanocomposites. The inclination angles of the faces with respect to the loading axis are 65.30 for Berkovich indenter. In the model, very fine meshes are selected using mesh sensitivity analysis so that Finite Element models can be simulated more accurately by providing minimum run time for models. Additionally, mesh biasness is also considered, as more numbers of elements are chosen near the indenter tip to make the analysis more efficient and hence desired convergence of output is achieved. Both material non-linearity in the mechanical properties of CNT and matrices and geometric non-linearity for the elements are employed in the FE model simulation. The indenter is considered to be analytically rigid; moreover, surface to surface interaction between CNT and matrix is also considered to be friction for non-bonded interface where matrix is the master surface and CNT - the slave. In the non-bonded interface, thermal stress is considered as uniform radial pressure acting on the nanotube.

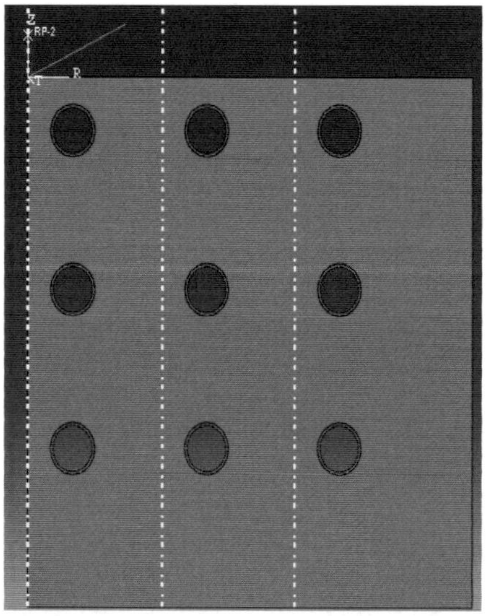

Figure 1. FE model for CNT in polymer composite SWCNT (without meshing).

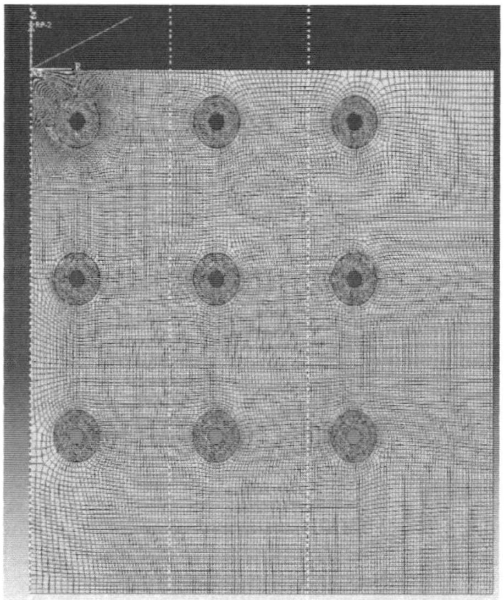

Figure 2. FE model for CNT MWCNT reinforced nanocomposite.

3. General equations for indentation relationships

In order to obtain the mechanical properties of the nanocomposites, the results are extracted from the FE analysis that is finally processed as force displacement relationships of the indenter tip as shown in Figure 3. In the analysis process, it is assumed that the area of contact between the indenter and the specimen remains constant while the unloading stage begins. The mechanical properties such as hardness and modulus of elasticity are extracted from the force-displacement relation as per well-known procedure proposed by Doerner and Nix (1986).

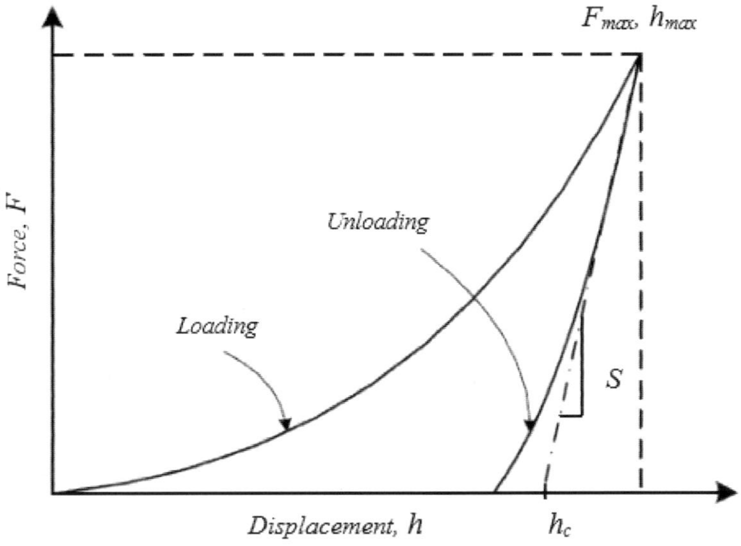

Figure 3. Schematic drawing for the determination of the material properties from indentation test proposed by Doener and Nix method (Doerner and Nix 1986).

The gradient of the Force-indentation displacement curve during unloading is given by the following relation

$$S = \frac{|dF|}{|dh|_{h=h_i}} = \mu\, E_{ef}\sqrt{A_p} \tag{1}$$

Where,

S	represents contact stiffness
h_t	is the total indentation depth
A_p	the projected area
$\mu = 2\sqrt{\pi}$	for axisymmetric indenter
$\mu = 1.167$	for Berkovich indenter
E_{ef}	effective elastic modulus
E_i	indenter elastic modulus
E_f	elastic modulus of the composite

The relationship between the indentation and the projected area of contacts can be expressed through the following equation

$$A_p = kh_p^2 \tag{2}$$

where,

h_p	effective indentation depth
$k = 24.5$	for Berkovich &Vicker indentation

The hardness (H), being expressed as the ratio of maximum applied load to the projected contact area for Berkovich indenter that becomes equal to

$$H = \frac{F_{max}}{A_p} = \frac{F_{max}}{24.5\, h_p^2} \tag{3}$$

Using equation (2) and equation (3), the mechanical properties of nanocomposites such as modulus of elasticity and hardness can be obtained.

4. Finite Element Analysis

Many researchers have studied finite Element Modeling for indentation analysis. However, most of them consider perfect CNT/matrix interface and their models does not account the hollowness of the CNTs. The hollowness effect can be a very important factor when diameter to wall thickness ratio is significantly high. To the best of author's knowledge, a combination of length scale effect and non-bonded CNT/matrix interface for nanocomposite to investigate up to a fraction of nanometer scale has never been studied. Considering all practical factors of CNTs and their potentials matrices, a series of Finite Element model has been simulated to obtain the accurate indentation output and the influences of different composite parameters. The parameters used in the finite element model are presented in Table 1.

Table 1. Value of parameters.

Property Description	Value
Domain Size of the axisymmetric Model	10 nm × 10 nm
Diameter of CNT	1 nm
Indenter tip displacement	0.788 nm
Number of CNT in the domain	9
Bias Ratio for meshing	10
No of element in each dimension	100
Modulus of Elasticity of CNT, E_t	1000 GPa
Poisson's Ratio of CNT, ν_t	0.27
Modulus of Elasticity of Epoxy, Em	3000 MPa
Poisson's Ration of Epoxy, ν_m	0.35
Thickness of CNT	0.034 nm
Radial Stress due to thermal contraction	60 MPa
Coefficient of Friction, μ	0.25
ϵ	0.004656 ev$(1\text{ev} = 1.602\times10^{-19}$ j)
n_p	3.1×10^{28} / m^3
n_c	3.82×10^{19} / m^2
O_i	0.2 nm

4.1. Model Validation

Before going to discuss the study on wall thickness dependencies of CNT reinforced nanocomposites, the proposed model is validated with the experimental work conducted by Nouri et al. (2012). In the modeling process, Berkovich indenter having 65.30 indentation angle and a depth of 740 nm,

IOP Conf. Series: Materials Science and Engineering **839** (2020) 012008 doi:10.1088/1757-899X/839/1/012008

mechanical properties of aluminum and MWCNT are exactly followed as per their experimental indentation test.

Nanoindentation model for 2% (by vol.) MWCNT reinforced AL-2024 nanocomposite is created and compared with that of the experimental indentation test. The stress strain relation of AL-2024 is extracted from the previous study conducted by Bastawros et al. (2006). Figure 4 shows a comparison of the force displacement relationship between numerical and experimental indentation test for 2% MWCNT reinforced AL-2024 nanocomposite. Von mises stress distribution of the nanocomposite is presented on Figure 5. The stress distribution shows that MWCNT experiences larger stress than the adjacent matrix that justify the reinforcing potential of MWCNT. It is also observed from the stress distribution that there are breaks of stress flows once they reach the MWCNTs. The numerical result for the nanocomposite well agrees with the experimental result in both loading and unloading stage. Though the curves intersect each other in some indentation depths, they follows the same trends and similar material behavior. The minor differences between the results may arise from the considerations that the numerical model considers the uniform dispersions of CNT, the indenter as analytical rigid and perfect orientation. However, it's nearly impossible to achieve uniform dispersion, perfect rigid indenter with infinite stiffness and theoretical orientation of a nanocomposite that is experimentally tested. The correlation obtained by numerical analysis appears indistinctly uneven in some points that actually happens due to the account of material and geometric non-linearity as well as the vdW interactions at the non-bonded interfaces.

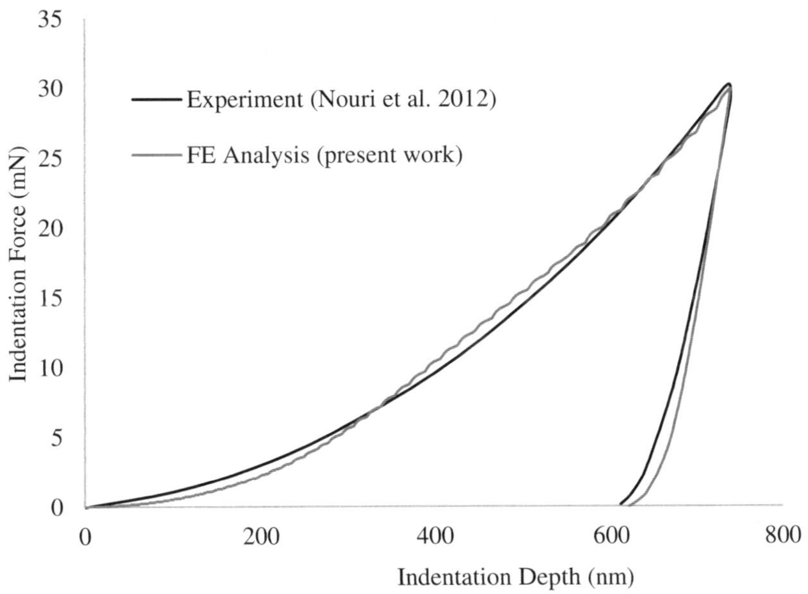

Figure 4. Schematic drawing for the determination of the material properties from indentation test proposed by Doener and Nix method (Doerner and Nix 1986).

IOP Conf. Series: Materials Science and Engineering **839** (2020) 012008 doi:10.1088/1757-899X/839/1/012008

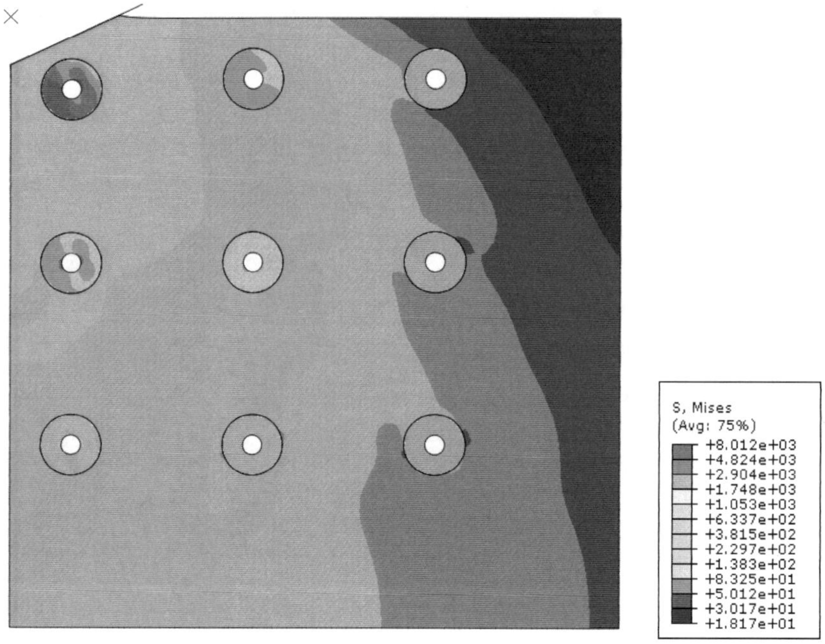

Figure 5. Stress distribution of nanoindentation simulation of MWCNT reinforced aluminum nanocomposite.

Table 2. Comparison between current study and previous experiment.

Parameter	Hardness (GPa)	Modulus of elasticity (GPa)
2% MWCNT reinforced Aluminum nanocomposite (Experiment (Nouri, Ziaei-Rad et al. 2012))	2.263	83.3
2% MWCNT reinforced Aluminum nanocomposite (present work)	2.20	82.6

5. Results and Discussions

Berkovich indentation analysis is performed on CNT reinforced conventional steel nanocomposites where the influence of different thicknesses are investigated. Analysis has been run for ten different wall thicknesses of CNTs keeping the same outer diameter. Elasto-plastic behavior of steel and CNT including strain gradient is considered in the program. Investigations are conducted in a wide range of wall thicknesses starting from 0.034 nm to 0.334 nm. The stress distribution of nanoindentation simulations for typical thin and thick walled CNTs are presented in figure 6a and 6b, respectively. The Berkovich simulation result shows that the higher stress is developed near the indenter tip. In addition, maximum von mises stress is observed in the CNTs compare to the adjacent matrix region. The stress distribution evidently shows that thin wall CNTs near the indenter are collapsed due to buckling inside the matrix. In contrast, thick wall CNT (having a diameter over 0.2 nm) seems to float inside the matrix without showing any buckling as indentation progresses. It is interesting to note that, though the indenter tip moves downward inside the matrix, the shape of the thick-walled tube remains unchanged. It is also observed from the stress distribution suggests that the stress flow is higher in lateral direction than that in vertical path.

Wall thickness dependency of mechanical properties of CNT reinforced nanocomposites is presented in terms of force-displacement relationship as shown in Figure 7. It can be clearly understood from the figure that force requirement increases as the indenter tip progresses through the nanocomposites.

However, the required force sharply increases for thicker nanotubes reinforced composites particularly after an indentation depth of 0.30 nm. On the other hand, the force increment of thinner walled (0.034 nm) CNT-Steel composite is not linear with the indentation depth. The difference of force requirement for same indentation depth between thin and thick walled nanocomposite is very much significant. The maximum required force of thick-walled CNT-steel composite for an indentation depth 0.788 nm is nearly 2.5 times than that of the thinnest CNT-Steel nanocomposites.

It is interesting to highlight that there is a turn at the force-displacement profile for an indentation depth in the range of 0.4 to 0.6 nm up to a wall thickness of 0.1 nm. This may happen due to the fact that buckling occurs for thin wall carbon nanotubes as presented on Figure 6a. This phenomenon can be explained by the fact that smaller wall thickness results local bucking of the CNTs, loses its stability and hence the resistance from the tube is insignificant in this region. However, the force requirement increases sharply after an indentation depth of more than 0.6 nm. This happens because though, the CNT loses its lateral stability after initial buckling, it regains stability after further buckling as they are confined by matrix materials and show high ductile behavior inside the matrix. It is to be noted that if wall thickness is more than 0.2 nm, all CNTs having similar outer diameter give almost same indentation profile. This can be explained by the fact that lateral buckling of CNT for that particular diameter can be resisted by a wall thickness of 0.2 (diameter to thickness ratio 5.0) or higher and this wall thickness may be considered as the optimum thickness to achieve maximum influence of CNT in nanocomposites for a fixed outer diameter of 1 nm.

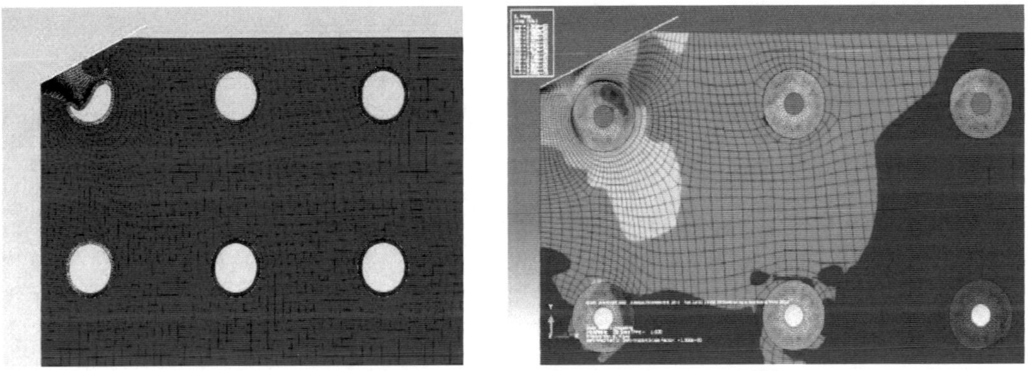

(a) (b)

Figure 6. Stress distribution of Berkovich indentation simulation of (a) thin-walled and (b) thick-walled nanotube reinforced composites.

Figure 8 shows the variation of modulus of Elasticity of nanocomposites (Enc) with the wall thickness of CNTs. The values of modulus of elasticity of nanocomposite are extracted from the force displacement relationships of wall thickness dependency study. The result shows that with the increase of wall thickness of CNT, modulus of elasticity of the composite also increases though the increment does not change linearly. Initially, the variation of Enc with respect to wall thickness is much steeper than that of the wall thickness greater than 0.2 nm. It is very worthwhile to mention that the modulus elasticity of CNT-steel nanocomposite for a wall thickness of less than 0.05nm (SWCNT) is observed to be nearly 180 GPa which is less than that of pure steel (Es = 200 GPa). Therefore, SWCNTs are found to be ineffective as a reinforcement of pure steel matrix.

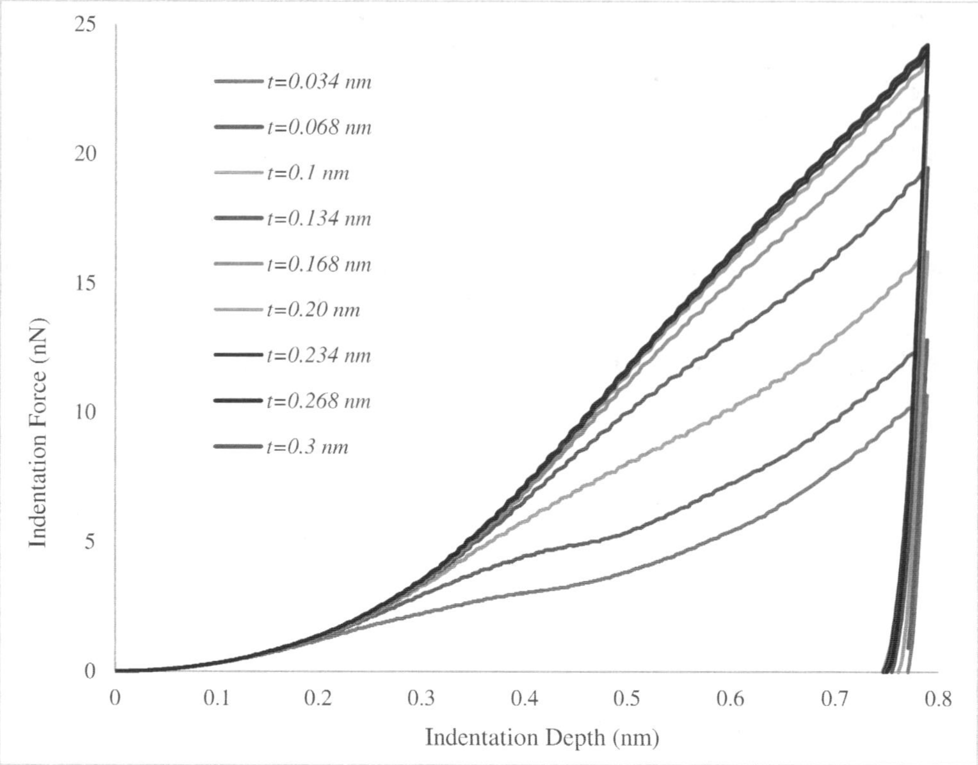

Figure 7. Force displacement relationship for different wall thicknesses of CNTs in nanocomposites.

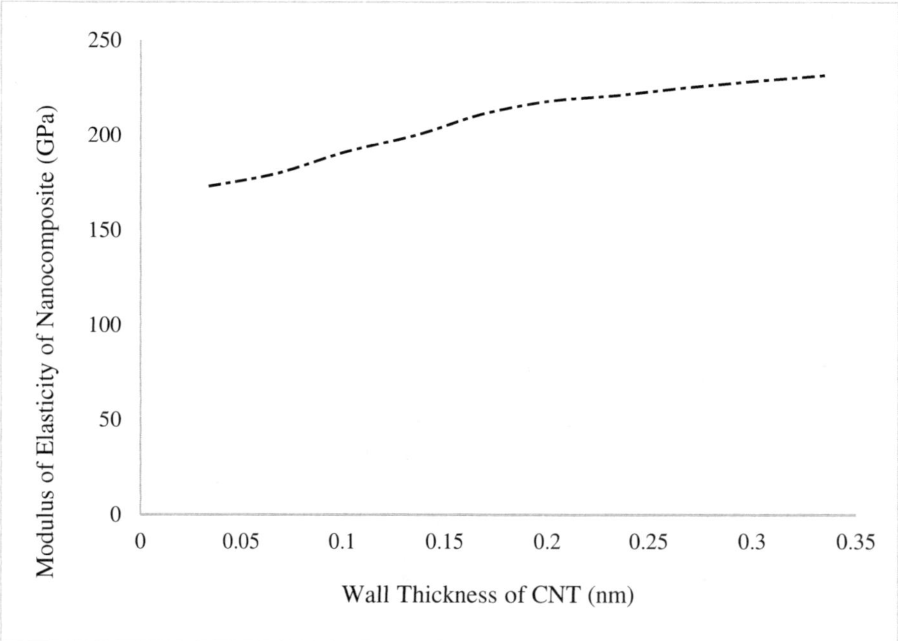

Figure 8. Influence of CNT's wall thickness on modulus of elasticity of nanocomposites.

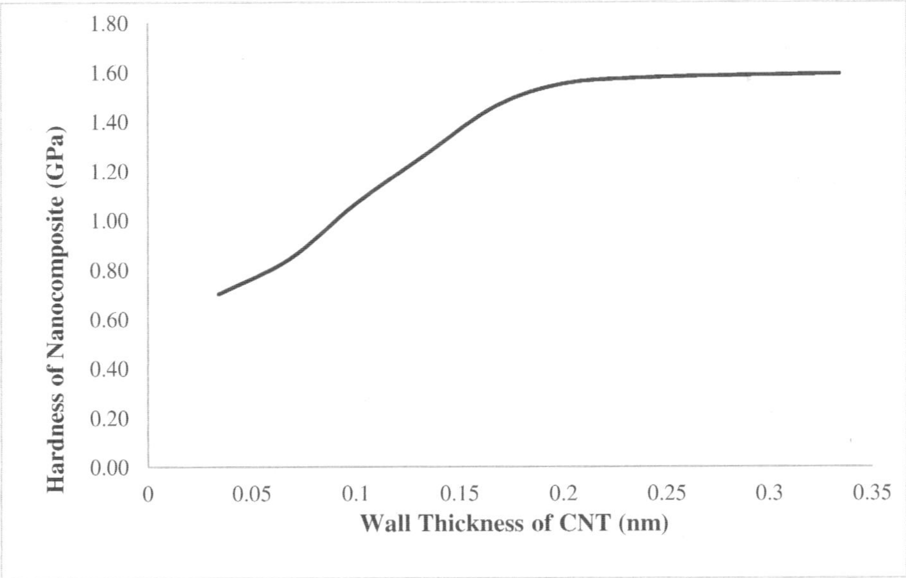

Figure 9. Influence of CNT's wall thickness on hardness of nanocomposites.

Figure 9 presents the wall thickness dependency of hardness of CNT reinforced nanocomposites. It is also observed from the figure that hardness sharply increases with the increase up to a wall thickness of 0.2 nm and it remains nearly constant after that value. This hardness behavior of nanocomposite can be described as, large distortion of CNT occurs for smaller wall thickness of CNTs. In addition, bond strength between collapsed CNT and adjacent matrix become weaker because cohesive stress due to vdW interactions becomes insignificant as the interface goes far away from the equilibrium distance. Therefore, in order to achieve a desired hardness for nanocomposite, this study may provide a preliminary guideline to design the optimum wall thickness for particular matrix so that CNTs can effectively contribute to the composite properties.

6. Conclusion

A number of finite element models have been developed to investigate indentation analysis for uniformly dispersed CNTs in different matrices in which non-bonded CNT/matrix interface and large strain elasto-plastic behavior of constituents are considered. The proposed finite element model for nanoindentation analysis is validated with the experimental indentation test on 2% (by vol.) MWCNT reinforced aluminum nanocomposites conducted by Nouri et al. (2012). The comparison shows that the proposed model demonstrates a very good agreement with their experimental result. The differences of modulus of elasticity and hardness between experimental indentation test and this advanced finite element simulation are below 1% and 3%, respectively.

The study on wall thickness dependencies of CNTs in nanocomposite shows that mechanical properties of nanocomposites largely depends on the wall thickness of CNTs. A minimum wall thickness for a particular tube diameter to achieve a pick value of hardness and modulus of elasticity of nanocomposites is determined. The minimum wall thickness of CNTs is suggested to be 0.2nm for an outer diameter of 1nm in order to achieve the maximum value composite properties. In addition, a wall thickness smaller than 0.05 nm may even reduce the hardness and modulus of elasticity of CNT reinforced composites compare to that of pure matrix. This happens due to the fact that large distortion occurs for smaller thickness. In addition, this distortion of CNTs changes the relative radial displacement at the CNT/matrix interface causing a smaller stress transfer through the interface and hence results smaller values of hardness and modulus of elasticity of nanocomposites. Therefore, this study may contribute in designing different type of carbon nanotube reinforced nanocomposites and their appropriate application as reinforcement of different nanostructures.

7. References

[1] Ajayan P M, Schadler L S, Giannaris C and Rubio A 2000 Single-Walled Carbon Nanotube–Polymer Composites: Strength and Weakness *J. Adv. Mater.* **12** 750-53

[2] Ashrafi B and Hubert P 2006 Modeling the elastic properties of carbon nanotube array/polymer composites *Compos. Sci. Technol.* **66** 387-96

[3] Bakshi S R, Lahiri D and Agarwal A 2010 Carbon nanotube reinforced metal matrix composites - a review *Int. Mater. Rev.* **55** 41-64

[4] Chen X 2004 Square representative volume elements for evaluating the effective material properties of carbon nanotube-based composites *Comput. Mater. Sci.* **29** 1-11

[5] Chen X H, Chen C, Xiao H N, Liu H B, Zhou L, Li S L and Zhang G 2006 Dry friction and wear characteristics of nickel/carbon nanotube electroless composite deposits *Tribol. Int.* **39** 22-28

[6] Manoharan M P, Sharma A, Desai A V, Haque M A, Bakis C E and Wang K W 2009 The interfacial strength of carbon nanofiber epoxy composite using single fiber pullout experiments *Nanotechnology* **20** 295701

[7] Ahmed K S and A K Keng 2013 Interface characteristics of nanorope reinforced polymer composites *Comput. Mech.* **52** 571-85

[8] Melick H G H V, Bressers O F J T, Toonder J M J D, Govaert L E and Meijer H E H 2000 A micro-indentation method for probing the craze-initiation stress in glassy polymers J. *Polym.* **44** 2481-91

[9] Fazio L D, Syngellakisa S, Wood R J K, Fugiuele F M and Sciumé G 2001 Nanoindentation of CVD diamond: comparison of an FE model with analytical and experimental data *Diam. and Relat. Mater.* **10** 765-769

[10] Kusano Y and Hutchings Y K 2003 Analysis of nano-indentation measurements on carbon nitride films *Surf. Coat.* **169** 739-742

[11] Larsson P, Giannakopoulos A E, SÖderlund E, Rowcliffe D J and Vestergaard R 1996 Analysis of Berkovich indentation *Int.J. Solids Struct.* **33** 221-248

[12] Doerner M F and Nix W D 1986 A method for interpreting the data from depth-sensing indentation instruments *J. Mater Res.* **1** 601-609

[13] Briscoe B J, Fiori L and Pelillo E 1998 Nano-indentation of polymeric surfaces *Journal of Physics D: Appl. Phys* **31** 2395

[14] Nouri N, Ziaei-Rad S, Adibi S and Karimzadeh F 2012 Fabrication and mechanical property prediction of carbon nanotube reinforced Aluminum nanocomposites *Mater. Design* **34** 1-14

[15] Bastawros A F 2006 Analysis of deformation-induced crack tip toughening in ductile single crystals by nano-indentation *Int.J. Solids Struct.* **43**(24) 7358-70

Tribological and physical performance of micro bearing concept for diluted epoxy filled with UHMWPE composite

Ahmad Aizuddin Ariffin[1], **Ahmad Zafir Romli**[2] **and Nik Noor Idayu Nik Ibrahim**[2]

[1]Faculty of Applied Sciences, Universiti Teknologi MARA, 40450 Shah Alam, Selangor, Malaysia
[2*]Centre of Polymer Composite Research and Technology (PoCResT), Institute of Science, Universiti Teknologi MARA, 40450 Shah Alam, Selangor, Malaysia
ahmad.aizuddin@outlook.com

Abstract. Short fillers have been widely used as reinforcing agents due to their successful effects on composite materials. However, studies for short fillers were mainly focusing on even dispersion of fillers in the composite system. As for the application related to bearing, the fillers needed to be deposited on the surface to improve the physical and tribological properties. Therefore, in this study, the fundamental of micro bearing concept was imparted by relating to the rheological study of polymer. These relations can be supported by the theory of particulate fillers in suspensions according to Modified Mooney-Einstein's Theory of Viscosity. The Ultra High Molecular Weight Polyethylene (UHMWPE) fillers, which possess high wear resistance, were fabricated with epoxy resin. The mixture was prepared by diluting with acetone solvent and varying the filler loadings percentage for 1%, 3%, 5%, 7%, and 10%, accordingly. Based on the results observed, there were increment up to 41.7% for viscosity value. For the wear test, the results showed decrease in the wear rate up to 72.0%. For the hardness test, the results indicate increment in the hardness value up to 94% compared to UHMWPE alone. For the density test, the results showed decrease in the density value up to 3.8%.

Keywords: Diluted epoxy, Micro bearing concept, Modified Mooney-Einstein's Theory of Viscosity, UHMWPE, Wear rate

1. Introduction

In definition, bearing is a system of machine element that restrict correlative motion between operating parts by providing support upon an applied force through reducing friction between the relatively moving surfaces [1]. As for the bearing applications, the improvement in the bearing capacity are demanding which need to be optimized by enhancing the durability of the bearing properties upon movement on the plane of products. Normally, bearing products consist of single bearing material alone which can be made from metals, ceramics or polymer materials. Recently, since polymer materials possess addition unique properties which are corrosion and chemical resistant, they are widely being selected as bearing materials for various types of contact applications [2]. Another bearing products consist of secondary phase on top surface whether in form of protective layer, sheet and lubricant to function as noise reduction, shock absorber and improved the fatigue life of the product [3]. However, the challenges for the existing bearing products, whether they consist of a single phase or secondary phase are the durability to function over time. This is due to that bearing product that is relatively too rigid which can break the load and cause high wear rate. For the bearing product that is too soft, it can cause potting and stress concentrator thus can disentangle the bearing surface upon creep and fatigue.

Therefore, in order to obtain the balanced tribological and physical properties for respective bearing surface, a study by Gahr [4] can be employed in this study whereas the findings show that the ruptured particles surface can be chemically modified by introducing a microstructure on the surface. The fundamental of polymer composite is being utilized whereas the bearing material which is Ultra High Molecular Weight Polyethylene (UHMWPE) polymer, in the form of micro-sized, has been chosen as the reinforcing fillers while epoxy resin has been selected as the matrix phase, consecutively. For UHMWPE, it is a semi-crystalline polymer with the dominant properties of high wear and abrasion resistance, chemical resistance and biocompatibility [5]. These properties makes it convenient for gears, liners and bearings application [6-7]. For epoxy resin, it consists of significant properties such as high solvent and chemical resistance, outstanding adhesion to substrates, low shrinkage upon curing, high flexibility and high impact resistance [7].

For the micro bearing concept, the fundamental study of material rheology which involves the Non-Newtonion fluid behaviour of polymer as relation to the viscosity and density study. In plastic material, when incorporating the short fillers, they will enhance the viscosity of the polymer mixture [8]. This will lead to the dilatant behaviour (shear thickening) of the Non-Newtonion fluid upon increase in the shear rate [9]. The theory that describe the behaviour of spherical fillers in polymers are first described by Einstein (1) for diluted solutions or dispersions. Einstein states that the viscosity of viscous fluid (η_o) increased when small, rigid and non-interacting spheres were suspended in the liquid [10]. According to the equation, the viscosity of the mixture (η) is directly proportional to the fractional volume (c) occupied by the particles which viscosity of the mixture does not depend on the polarity of the liquid and size of the spheres. The parameters are fractional volume (c) which is less than 0.1 and maximum packing of sphere is 90%. The introduction of a coefficient which is 2.5 indicates that it is for hard particle spheres. The equation has been modified by Mooney (2) by introducing hydrodynamics or crowding factor (β) which is equal to 1.35 and 1.91 for loosely and densely packed spheres. The modification can be similar to the original Einstein Equation when the crowding factor is equal to 0. The Einstein Equation and Modified Mooney-Einstein Equation can be deduced as follow.

$$\eta=\eta_o(1+2.5c) \quad \text{and} \quad \eta_{sp}/c=2.5 \tag{1}$$

$$\eta = \eta_o\ (2.5c/1-\beta c) \quad \text{and} \quad \eta_{rel}/c=2.5 \tag{2}$$

Since the increase in the filler loadings will contribute to the increase in the viscosity of the polymers, the need of diluent is significant to reduce the viscosity for the matrix. Hence, by introducing diluent, less dense filler can be deposited on the surface of composite upon curing. The methodology is different from the normal composite system which need the fillers to be evenly distributed inside the composite system. As for the diluent, acetone solvent has been widely used by various studies to improve the processing parameter and behaviour of respective fillers inside the matrix constituents. Based on various studies conducted by Gong *et al.*, Loos *et al.* and Dong *et al.* [11–13], the outcomes show that acetone solvent reduce the viscosity of the resin, which then will allow a better dispersion and distribution of fillers upon curing.

The aim of this work is to study the effect of filler loadings on the tribological and physical performance of the micro bearing concept by fabricating the diluted epoxy filled with UHMWPE as a composite system. The mixture of epoxy resin were successfully prepared by diluting with acetone diluent and varying the UHMWPE filler loadings at different weight percentage. During processing, the viscosity of the mixtures were tested using viscosity test before being proceed on the tribological and physical tests upon curing. The tribological test, which is the wear test was conducted to determine the wear rate which can be deduced as [14],

$$W=\Delta V/F_N x\ L \tag{3}$$

whereas W is the specific wear rate, ΔV is the volume loss (mm^3), F_N is the normal load (N), and L the sliding distance (m). The morphological analysis after wear was also conducted under optical microscopy measurement using Polarized Optical Microscope to observe the nature of sample after abrasion. For the physical tests, hardness test was conducted to determine the hardness value while density test was conducted to measure the density value of the respective samples.

2. Methodology

2.1. Materials

The Epoxy Resin, which consists of Epoxen CP362 'Part A' Resin (Epoxy DGEBA type) and Epoxen CP362 'Part B' hardener (modified aliphatic amine) were purchased from Oriental Option Sdn. Bhd. The UHMWPE GUR 4120 fillers were supplied from Ticona Engineering Polymer, China in powdered form with molecular weight of 5 x 10^6 g/mol and density of 0.93 g/cm^3. The AR1003-P2.5L acetone, which meet the A.C.S Specification, with molecular weight of 58.08 g/mol and density of 0.79 g/cm^3 at 25 °C was obtained by RCI Labscan Limited Company.

2.2. Samples Preparation

The micro bearing concept of Epoxy filled UHMWPE composites were mixed and fabricate using casting technique. The methodology of preparation the samples were based on study of Ariffin *et. al.* [15] with some modifications to obtain the micro bearing layers. First, the epoxy resin was stirred with UHMWPE fillers using a mechanical stirrer at 650 rpm for 10 minutes followed by sonication process at 20 kHz for 10 mins. Then, the mixture were diluted with acetone solvent for 5 minutes at 250 rpm. After that, the mixture were stirred with hardener for 5 mins at 250 rpm. The diluted mixture were later being hold at ambient temperature for 5 minutes before being poured into an open mould. Later, the samples were cured at room temperature for 24 hours before undergo post-cure process in the oven at 50 °C for 24 hours. Upon curing, the micro bearing layers can be observed with the distinction of the two phases on the composite sample. The deposition of fillers on the surface of composite can be illustrated based on Scanning Electron Microscopy analysis as shown on Figure 1. In this study, the research variables were the weight percentage of UHMWPE filler loadings in the composite system which are 1% (EpUPE1), 3% (EpUPE2), 5% (EpUPE3), 7% (EpUPE4) and 10% (EpUPE5), accordingly. While for the control sample, neat epoxy (Ep) was used as reference sample.

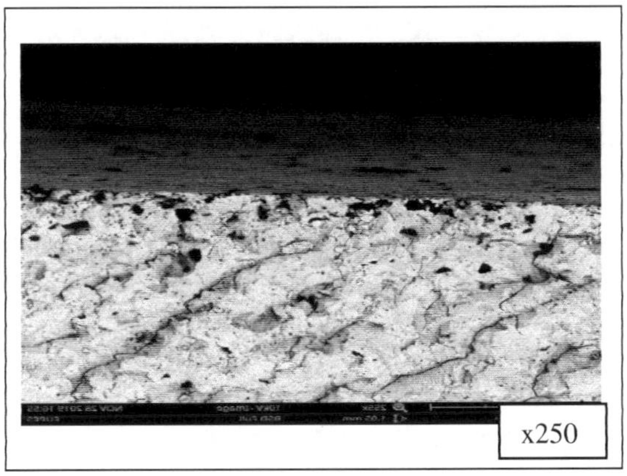

Figure 1. The filler disposition on the surface of micro bearing layers under Scanning Electron Microscopy analysis (Source by author).

2.3. Experimental Testing

IOP Conf. Series: Materials Science and Engineering **839** (2020) 012009 doi:10.1088/1757-899X/839/1/012009

2.3.1. Viscosity Test.
The viscosity test was conducted using a Brookfield DV-1 Prime Viscometer of spindle fiber number 0.2 with rotating speed of 2.5 rpm. The measured volume of 200 ml polymer resin in liquid form was placed in a cup and the viscosity reading was taken continuously every minute for a total of 20 minutes. The results were observed in viscosity unit (mPas) against time (seconds). The test procedure was conducted according to ASTM Standards D445-06 [16].

2.3.2. Wear Test.
The wear test was conducted using Sangyo Co. Ltd. AB-301 Color Fastness Rubbing Tester with 6 N load and 0.01m sliding distance. The pin configuration was a Pin-on-Table type with 5000 cycles in 3 hours. The results were measured in mm^3/Nm. The test procedure was conducted according to ASTM Standards G132-96 [17].

2.3.3. Hardness Test.
The hardness test was examined using an Instron Wilson/Rockwell A654-R Digital Hardness Tester with 60N major load and 5 seconds delay time. The ball indenter diameter was 12.7 mm which is in the scale of 1:1/16. The results were measured in Rockwell Hardness Scale R (HRR). The test procedure was conducted according to ASTM Standards D-785 [18].

2.3.4. Density Test.
The density test was conducted using an Electronic Densimeter MD-300S with $0.001 g/cm^3$ density resolution. The specimens were first weighed on top of the densimeter before being immersed in the densimeter water container to determine the apparent mass. After that, the specific gravity in term of relative density were calculated. The test procedure was conducted according to ASTM Standards D792-17 [19].

2.3.5. Optical Microscopy Measurements.
The optical microscopy measurements were taken by using Olympus BX51 Polarized Reflected Microscope under bright field mode with U-TV1X-2 MoticamPro Camera Adapter. The observations were carried out under 5x magnification with controlled fluorescent ring illumination. The test procedure was conducted according to ASTM Standards E883-11 (2017) [20] .

3. Results and Discussions

3.1. Viscosity Test
For rheological study, the kinematic viscosities of the composite mixture with difference UHMWPE filler loadings can be seen on Figure 2. From the graph, the trend shows that incorporating filler loadings enhance the viscosity of the mixture. EpUPE5 shows the highest range of viscosity value which are around 17000 mPAs while EpUPE1 shows the lowest range of viscosity value which are around 12000 mPAs. This indicates that there is up to 41.7% viscosity increment from low filler loadings (1%) to high filler loadings (10%), respectively.

IOP Conf. Series: Materials Science and Engineering **839** (2020) 012009 doi:10.1088/1757-899X/839/1/012009

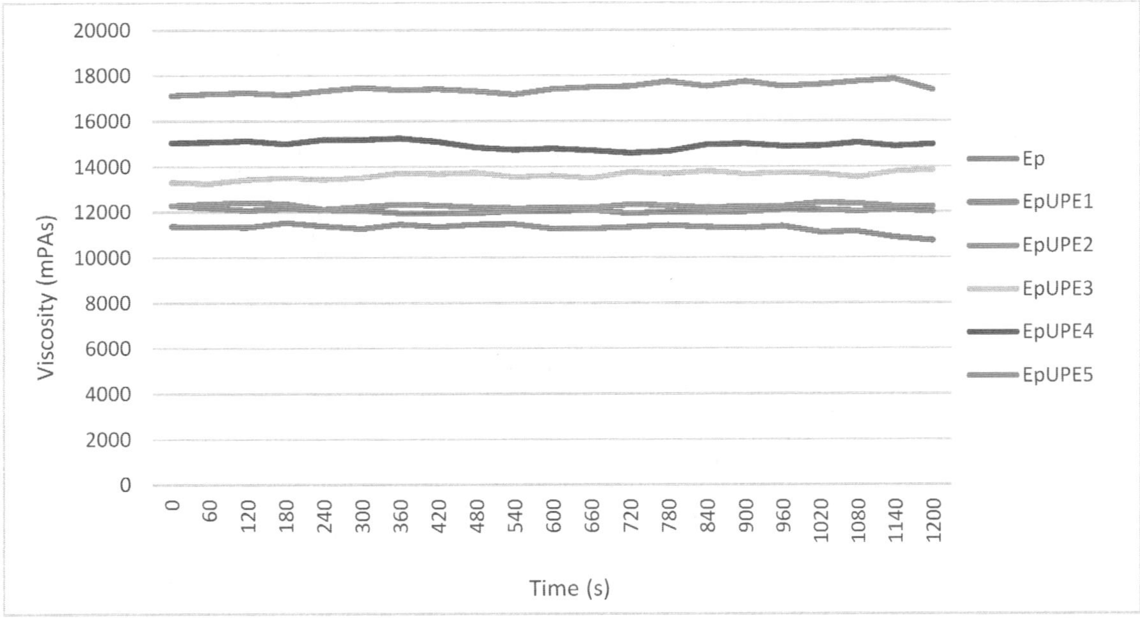

Figure 2. The kinematic viscosities of difference filler loadings against time.

For the Ep mixture which act as reference sample, the range of viscosity of the neat epoxy resin is around 10000-11000 mPas indicates that the matrix is originally in high viscous flow behavior. Upon incorporating UHMWPE fillers into the mixture, the viscosity value increase significantly proved that it follows dilatant behaviour of polymer flow [9]. This is due to the UHMWPE particles tend to form agglomeration inside the matrix thus resulting restriction of material flow. The increase in the resistance of the flow makes more force needed to flow the mixture upon rotational motion as relation to the fundamental of viscosity [21]. Furthermore, since the fillers used are short fillers, they can penetrate randomly through the mixture of epoxy resin which provide the interfacial filler-matrix adhesion of the UHMWPE particles with the resin. The data obtained also had been proven based on the Modified Mooney-Einstein Theory of Viscosity whereas viscosity of the overall composite will be increased linearly upon increase in the filler loadings.

Hence, with the relatively high viscosity of composite mixture with increase in filler loadings, the micro bearing concept cannot be incorporated as the filler as they will randomly dispersed into the composite system upon curing. However, acetone diluent has been used to reduce the viscosity of the mixture so that the distinctive density of epoxy resin and UHMWPE filler can be separated into two phases to form the micro bearing layers. The micro bearing layers formation with acetone diluent and the epoxy filled with UHMWPE fillers without acetone can both be illustrated on Figure 3a and Figure 3b.

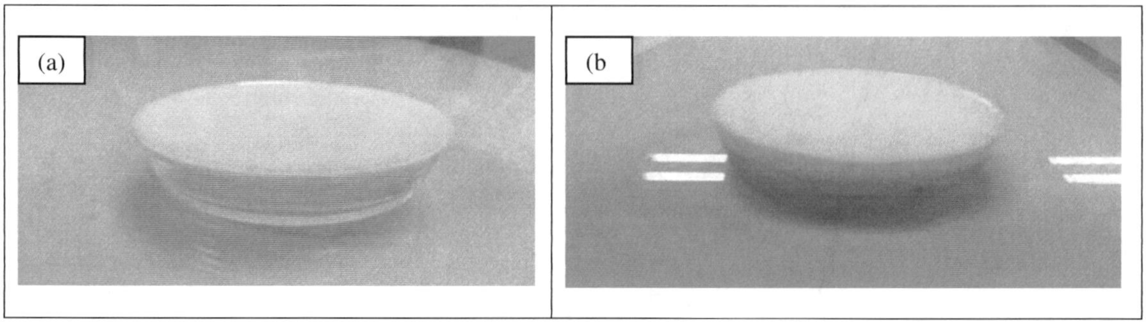

Figure 3. a) The micro bearing layers formation with acetone diluent and, **b)** The epoxy filled UHMWPE fillers without acetone diluent (Source by author).

3.2. Wear Test

For wear test, the wear rate values of the composite with different filler loadings can be tabulated based on Figure 4. From the graph, the highest wear rate value is UHMWPE1 with 3.22×10^{-5} mm³/Nm while the lowest wear rate value is 0.9×10^{-5} mm³/Nm. The trend show a decline in wear rate with increase in the filler loadings. For the Ep sample, it can be said that it slightly increase the wear rate value due to the nature of epoxy which is brittle upon curing which is subject to catastrophic failure upon loading [7]. When incorporating filler from low to high loadings, the wear rate value decreases. This can be explained by the nature of the UHMWPE polymer which is commonly known for its dominant properties of improved the wear resistance and reducing the wear rate [22]. When incorporating an increase amount of UHMWPE filler, the surface of the bearing layers will be deposited by the aggregation of these fillers. The clump of fillers will somehow act as a cushion which reduce the brittleness of the matrix constituent. As a result, the stress loading behaviour can be transferred and distributed between the fillers thus reducing the friction and impact of load upon motion. Therefore, volumetric loss can be minimize when two surface being contacted with each other for period of time as referring to the wear rate equation [14].

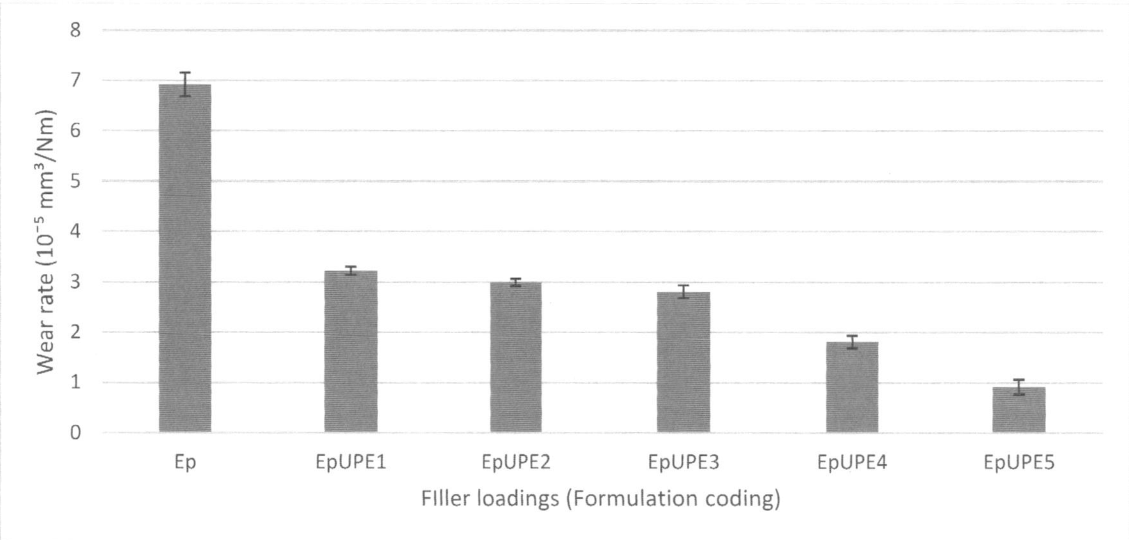

Figure 4. The wear rate values of the composite with different filler loadings.

The morphological after wear for all of the respective samples were conducted as a relation to the data obtained from the study. The abrasive phase as well as the worn surface were analysed by using optical microscopy measurements on the surface of the micro bearing layers.

3.3. Hardness Test

The effect of UHMWPE filler loading on the hardness value of the composites were presented in column diagram as shown in Figure 5. As comparison, EUHMWPE5 yield the highest hardness value with 97 HRR while EUHMWPE10 yield the least hardness 88 HRR. The trend of the graph show that increasing in the filler loading concentrations will tend to increase the hardness value until EUHMWPE5 and significantly drop the value from EYHMWPE 7 to EUHMWPE10. Based on the data observed, it can be said that incorporation of fillers give significant effect depending on several factors like density, distribution and dispersion of the fillers themselves inside the composite matrix.

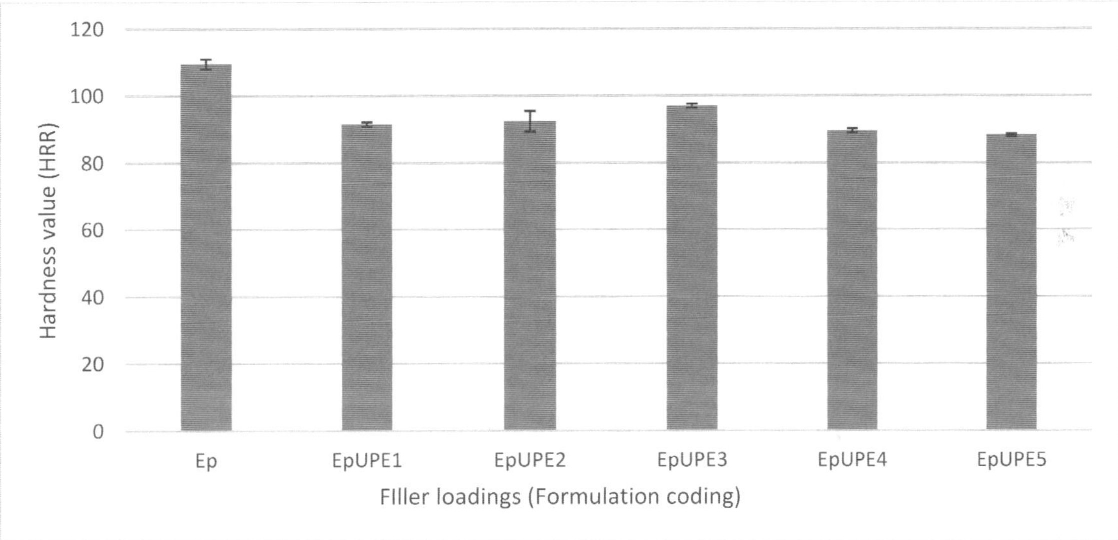

Figure 5. The hardness value of the composite different filler concentrations.

Generally, for Ep sample tend to generate high hardness value which is 110 HRR. Upon mixing with acetone and UHMWPE fillers, the values drop indicates that acetone reduce the brittleness of the epoxy matrix and provide slightly elastic behaviour of the composite [23]. The slightly decrease in the hardness value with filler loadings are tolerable since the range hardness values, which are between 88 to 97 HRR, are way more harder and stronger than typical hardness value of UHMWPE alone which is 50 HRR [24]. As referring to the trend on the graph, for low filler loadings (EUHMWPE1, EUHWMPE3 and EUHMWPE5), it can be said that upon increase the aggregation of UHMWPE fillers into the epoxy matrix, they will provide a relatively packed region on the surface thus providing stronger phase to resist deformation. However, starting on high filler loadings (EUHMWPE7 and EUHMWPE10) onwards, the hardness values are slightly decline. The reason may be due that the filler-filler interaction which has reach the threshold value and being saturate on the surface. With the filler-filler interaction being the dominant phase compared to filler-matrix interaction, there might be lack of adhesion between fillers thus softening the surface when being clump together on a large scale. The data can be further supported by the relation with the density test and optical microscopy measurements of the filler loadings distribution on the surface of the micro bearing layers.

3.4. Density Test

For density test, the relationship between density value with filler loadings can be shown on Figure 6. From the graph, EUHMWPE1 yield higher density value with 1.137 g/cm³ while EUHMWPE5 yield the least density value with 1.095 g/cm³. For Ep sample, the density of the epoxy resin is relatively higher due to nature of epoxy resin that possess high viscosity and has low shrinkage upon cure [7]. This will lead to reduce in the void formation inside the matrix when curing. For the sample with filler loadings, the trend show that the density value of the composite decrease with increase in the filler loadings.

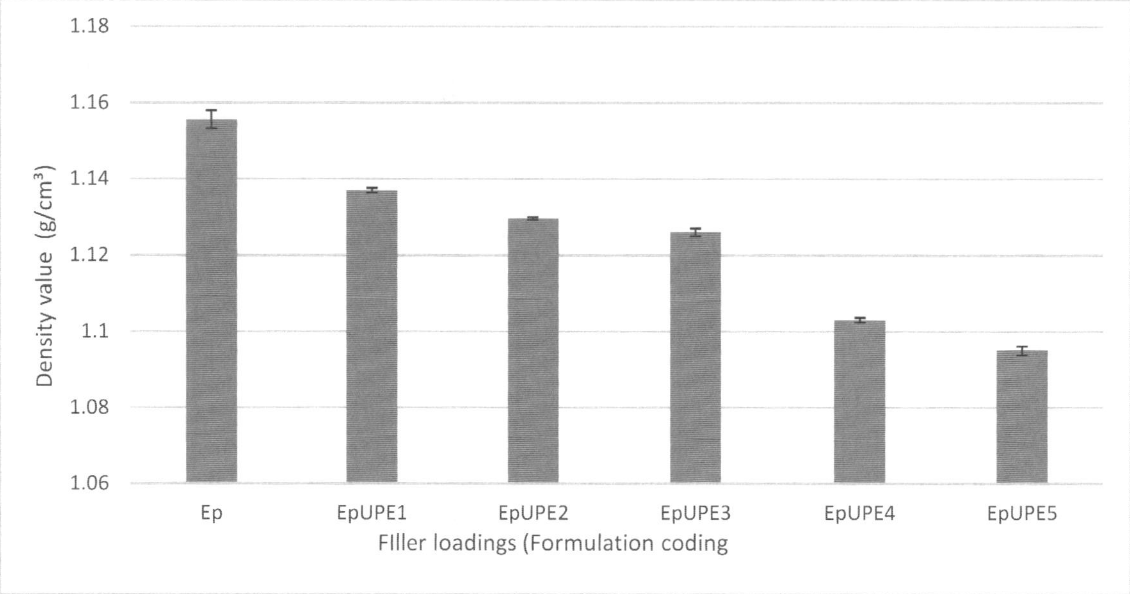

Figure 6. The density test for the respective formulation coding with difference filler loadings.

Based on the study conducted by Sudheer et al, hardness test is directly proportional to the density value [25]. The outcomes showed that the filler being used which is PTW fillers is harder and denser than the matrix which yield the dominant density and hardness properties upon increase in the filler loadings, respectively. However, the correlation are difference for this study wheras the UHMWPE filler being used are less dense filler with density of 0.96 g/cm³. Comparatively, when imparting a large amount of these fillers with specific amount of acetone, it will slightly reduce the overall viscosity of the composite which make the fillers to be the dominant phase on the surface of the micro bearing layers. The fillers that being clumped together on the surface will somehow make the composite to yield the density value closing to the dominant phase, which are the fillers as compared to the matrix phase.

3.5. Optical Microscopy Measurements

3.5.1. Wear Test.

For the wear test, the morphological analysis of the worn surface of Ep sample, EpUPE sample with low filler loadings and high filler loadings can be illustrated on Figure 7. Based on the optical microscopy measurements, all of the respective samples possess abrasion pattern perpendicular to the sliding direction of the pin with the same load cycles. The abrasion pattern can be represented on the circle of the respective samples. For Ep sample, in Figure 7a, the surface possess a high worn and abrasion upon contacting with the load. Upon the lateral direction of the sliding load, Ep sample show a wear debris in the form of black particles indicates that the surface has been worn out and the surface undergo plastic deformation. For low filler loadings, in Figure 7b, it shows that the abrasion pattern is reducing indicates

IOP Conf. Series: Materials Science and Engineering **839** (2020) 012009 doi:10.1088/1757-899X/839/1/012009

that UHMWPE filler provide slightly abrasion resistance by reducing the load friction thus restrict the high impact cycle load upon motion. For high filler loadings, in Figure 6c, the abrasion pattern show the least pattern due to the crumple formation of UHMWPE fillers on the surface. They will tend to form the surface cushion that which act as 'roller' that absorb and transfer the load between fillers into the micro bearing layers surface.

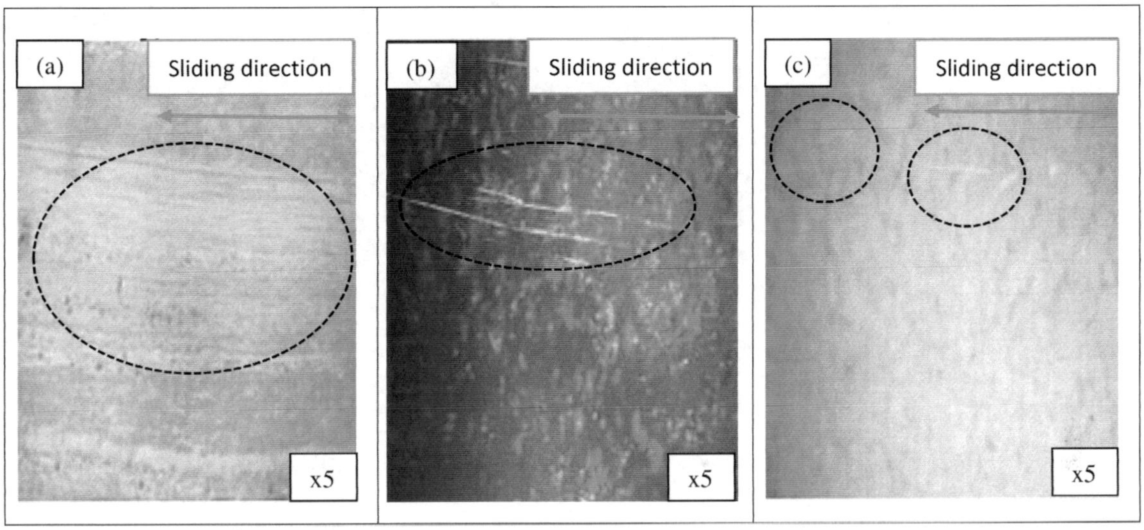

Figure 7. The optical microscopy of worn surface of **a)** Ep sample, **b)** EpUPE sample (low filler loadings) and, **c)** (high filler loadings) (Source by author).

3.5.2. Hardness Test

Hardness test can be supported by the morphological analysis of Ep sample, EpUPE sample with low filler loadings and high filler loadings, as shown on Figure 8. From the optical measurements, the figures show that UHMWPE fillers are generally being deposited on the surface of the composite as a clump of micro bearing layers. For Ep sample, in Figure 8a, the surface of the neat epoxy is shiny and tend to become glassy-like surface. This is due to the nature of epoxy which is hard but brittle in structure. For low filler loadings, in Figure 8b, it shows that epoxy resin fills the space between UHMWPE fillers which act as adhesion to strengthen the filler-matrix interaction on the surface. For high filler loadings, in Figure 8c, the crumple of fillers form filler-filler interaction which dominate the epoxy on the surface encapsulation layer thus represent the medium to resist deformation.

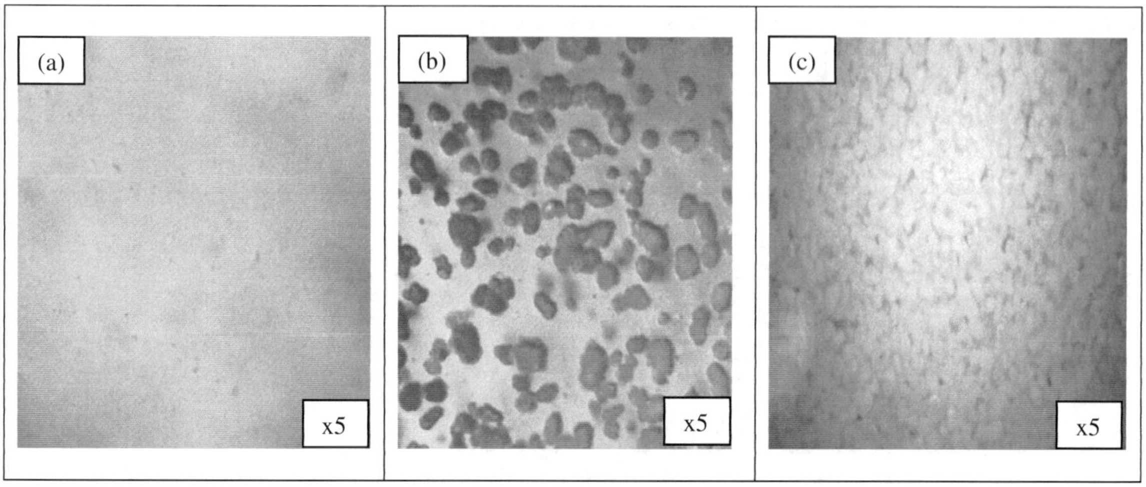

Figure 8. The optical microscopy for micro bearing layers surface of **a)** Ep sample **b)**, EpUPE sample (low filler loadings) and, **c)** (high filler loadings) (Source by author).

4. Conclusion

The effect of filler loadings on the tribological and physical performance of the micro bearing concept by fabricating the diluted epoxy filled with UHMWPE has been studied. The mixture of epoxy resin were successfully prepared by diluting with acetone diluent and varying the UHMWPE filler loadings at different weight percentage. Based on the results observed, EpUPE5 shows the highest range of viscosity value which are around 17000 mPAs while EpUPE1 shows the lowest range of viscosity value which are around 12000 mPAs. This indicates that there is up to 41.7% viscosity increment. For the wear test, the highest wear rate value is UHMWPE1 with 3.22×10^{-5} mm³/Nm while the lowest wear rate value is 0.9×10^{-5} mm³/Nm which indicates that it decreases the wear rate up to 72.0%. For the hardness test, EUHMWPE5 yield the highest hardness value with 97 HRR while EUHMWPE10 yield the least hardness 88 HRR, indicates that increment in the hardness value compared to UHMWPE alone up to 94%. For the density test, EUHMWPE1 yield higher density value with 1.137 g/cm³ while EUHMWPE5 yield the least density value with 1.095 g/cm³. The overall value shows that filler loadings decreases the denisty value up to 3.8%. The improvements on all of these properties are expected to be the important parameters for the utilization of micro bearing concept to counter the issues related to bearing applications especially on tribological and physical properties of the micro bearing layers of the composite.

5. References

[1] Majumdar B C 2008 *Introduction to Tribology of Bearings* (New Delhi: S. Chand & Company Ltd.) p 298
[2] Rubens N J, Seshagiri R S and Harihara S S P 2013 A review on the materials used for bearing and failure behavior *Int. J. of Stud. Res. in Technol. & Mgmt.* **1** 431-443
[3] Qiu M, Chen L, Li Y and Yan J 2017 *Bearing Tribology: Principles and Applications* (Berlin/Heidelberg: Springer-Verlag) p 333
[4] Gahr K H Z 1987 *Microstructure and Wear of Materials* (Amsterdam: Elsevier) p 560
[5] Rocha L F M, Cordeiro S B, Ferreira L C, Ramos F J H and Marques M F 2016 Effect of carbon fillers in ultrahigh molecular weight polyethylene matrix prepared by twin-screw extrusion *Mater. Sci. Appl.* **7** 863-880
[6] Puértolas J A and Kurtz S M 2016 UHMWPE matrix composites *UHMWPE Biomaterials Handbook (Third Edition)* ed S M Kurtz (Amsterdam: Elsevier) pp 369-397
[7] Massingill Jr J L and Bauer R S 2000 Epoxy resins *Applied Polymer Science: 21st Century*

IOP Conf. Series: Materials Science and Engineering **839** (2020) 012009 doi:10.1088/1757-899X/839/1/012009

ed C D Craver and C E Carraher Jr (Amsterdam: Elsevier) pp 393-424

[8] George K E 2009 Non-Newtonian fluid mechanics and polymer rheology *Advances in Polymer Processing: From Macro To Nano Scales* ed S Thomas and Y Weimin (Cambridge: Woodhead Publishing) pp 13–46

[9] Silberberg A 1977 Basic rheological concepts *Mucus in Health and Disease (Advances in Experimental Medicine and Biology* vol 89*)* ed M Elstein and D V Parke (Boston: Springer) pp 181-190

[10] Carraher J C E 2003 *Seymour/Carraher's Polymer Chemistry: Sixth Edition* (New York: Marcel Dekker, Inc.) p 960

[11] Gong X, Liu J, Baskaran S, Voise R D and Young J S 2000 Surfactant-assisted processing of carbon nanotube/polymer composites *Chem. of Mater.* **21** 1049-1052

[12] Loos M R, Coelho L A F, Pezzin S H and Amico S C 2008 The effect of acetone addition on the properties of epoxy *Polímeros: Ciência e Tecnol.* **18** 76-80

[13] Dong B, Yang Z and Li H L 2005 Study on tribological properties of multi-walled carbon nanotubes/epoxy resin nanocomposites *Tribol. Lett.* **20** 251-254

[14] Wang Y, Yin Z, Li H, Gao G and Zhang X 2017 Friction and wear characteristics of ultrahigh molecular weight polyethylene (UHMWPE) composites containing glass fibers and carbon fibers under dry and water-lubricated conditions *Wear* **380–381** 42–51

[15] Ariffin A A, Ibrahim N N I N, Shuib S, Romli A Z and Ismail N F 2019 The effect of acetone dilution towards the surface topography and morphology of micro bearing concept for epoxy filled UHMWPE composite *Sci. Res. J.* **16(2)** 27–42

[16] ASTM D445-06 2006 Standard test method for kinematic viscosity of transparent and opaque liquids (and calculation of dynamic viscosity) *ASTM International*

[17] ASTM G132-96 2018 Standard test method for pin abrasion testing *ASTM International*

[18] ASTM D785-08 2015 Standard test method for rockwell hardness of plastics and electrical insulating materials *ASTM International*

[19] ASTM D792-13 2013 Standard test methods for density and specific gravity (relative density) of plastics by displacement *ASTM International*

[20] ASTM E883-11 2017 Standard guide for reflected-light photomicrography *ASTM International*

[21] Viswanath D S, Ghosh T K, Prasad D H L, Dutt N V K and Rani K Y 2007 *Viscosity of Liquids: Theory, Estimation, Experiment, and Data* (Netherlands: Springer) p 662

[22] Selyutin G E, Gavrilov Y Y, Voskresenskaya E N, Zakhrov V A, Nikitin V E and Poluboyarov V A 2010 Composite materials based on ultra high molecular polyethylene: properties, application prospects *Chem. Sustain Dev.* **18** 301-314

[23] Abbas L K 2011 Study of the effect of acetone dilution percentage on mechanical properties of epoxy-acetone system *Iraqi J. Mech. Mater. Eng.* **11** 371-379

[24] Tiwari A, Murugan N A and Ahuja R 2016 *Advanced Engineering Materials and Modeling* (USA: Wiley-Scrivener Publishing) p 528

[25] Sudheer M, Prabhu R, Raju K and Bhat T 2014 Effect of filler content on the performance of epoxy/PTW composites *Adv. Mater. Sci. Eng.* **2014** 1-11

Acknowledgements
The authors would like to express endless appreciation to the Institute of Research Management & Innovation (IRMI), UiTM for the financial aid by Ministry of Education (MOE), Malaysia under the Fundamental Research Grant Scheme (FRGS), (600-IRMI/FRGS 5/3 (114/2019)).

IOP Conf. Series: Materials Science and Engineering **839** (2020) 012010 doi:10.1088/1757-899X/839/1/012010

Modeling of cleaning up forest debris technology from wood natural mortality

S P Karpachev[1], A A Shadrin[1], O N Burmistrova[2] and R I Diev[1]

[1] Mytishchi branch of Moscow State Technical University named after N. E. Bauman, 1, 1st Institutskaya street, Moscow region, the city of Mytishchi, 141005, Russia
[2]Ukhta State Technical University, 13, Pervomaiskaya street, Ukhta, Komi Republic, 169300, Russia

Abstract. The article considers the problem of cleaning up forest debris caused by fallen dead wood or wood natural mortality (WNM). Solution to the problem of cleaning and utilization of the WNM and quality control of these sanitary measures in forest was proposed. For conditions of "Medynskiy forestry" enterprise it was proposed technological process of cleaning forest from the WNM. A mathematical model and a conceptual model of the technological process were developed. The maximum values of the utilization coefficients of the winch and the feller are achieved at 30-meter length of the rope (Figure 8). With the increase in the volume of the WNM, the utilization coefficient of the winch increases, and the utilization coefficient of the feller are falls (Figure 9). However, these changes are insignificant (no more than 3.5%) and may not be taken into account in practice. Therefore, for the conditions adopted in the experiments, the length of the traction rope equal to 30 m can be considered optimal for portable hand winches.

1. Introduction

The article considers the problem of cleaning up forest debris and whole trees caused by fallen dead wood or wood natural mortality (WNM) (Figure 1).

The volume of the WNM in urban forests of Central Russia is estimated from 5 to 15% of the wood stock. According to the Committee of forestry of the Moscow region only in the forests of the Moscow region has accumulated about 20 million cubic meters of the WNM. Urban forests are used not only for logging, but, above all, they perform sanitary and health functions, are protective strips along the railways and roads. In such forests, cleaning of the WNM for aesthetic, fire protection and phytoprotection purposes, as well as for the safety of visitors should be carried out regularly. Cleaning of the forest from the WNM is regulated by sanitary rules and is appointed according to the results of forest pathology examination. One of the main objectives of these sanitary measures is to clear the forest from the WNM.

The existing technologies of cleaning of the forest from the WNM are imperfect, based on the use of manual labor. This article discusses an approach to the development of technology for harvesting and utilization of the WNM based on logging machines and their systems, which are used in low-volume logging.

Figure 1. The WNM in the forest in the form of whole fallen trees (photo by the author).

2. General approaches to cleaning forest from the WNM the and recycling of the WNM

Schematically, the solution of the tasks of cleaning forest from the WNM andrecycling of the WNM is presented in the form of a block diagram (Figure 2), consisting of four stages.

The first and last stages can be referred to the preparatory and final works. The purpose of the first stage is the justification and appointment of sanitary measures for cleaning of the forest from the WNM. At this stage, the scope, quality (for example, the percentage of commercial timber) and the distribution of the WNM on the forest plot. At the last stage, the quality of the forest cleaning from the WNM is monitored.

With data on the volume of the WNM, the second and third stages are formed systems of machines for cleaning forest from the WNM and recycling of the WNM. It should be noted that the volume of the WNM is usually many times less than the forest fund on the same plot. Therefore, the use of conventional logging machines for cleaning forest from the WNM is not economically feasible.

IOP Conf. Series: Materials Science and Engineering **839** (2020) 012010 doi:10.1088/1757-899X/839/1/012010

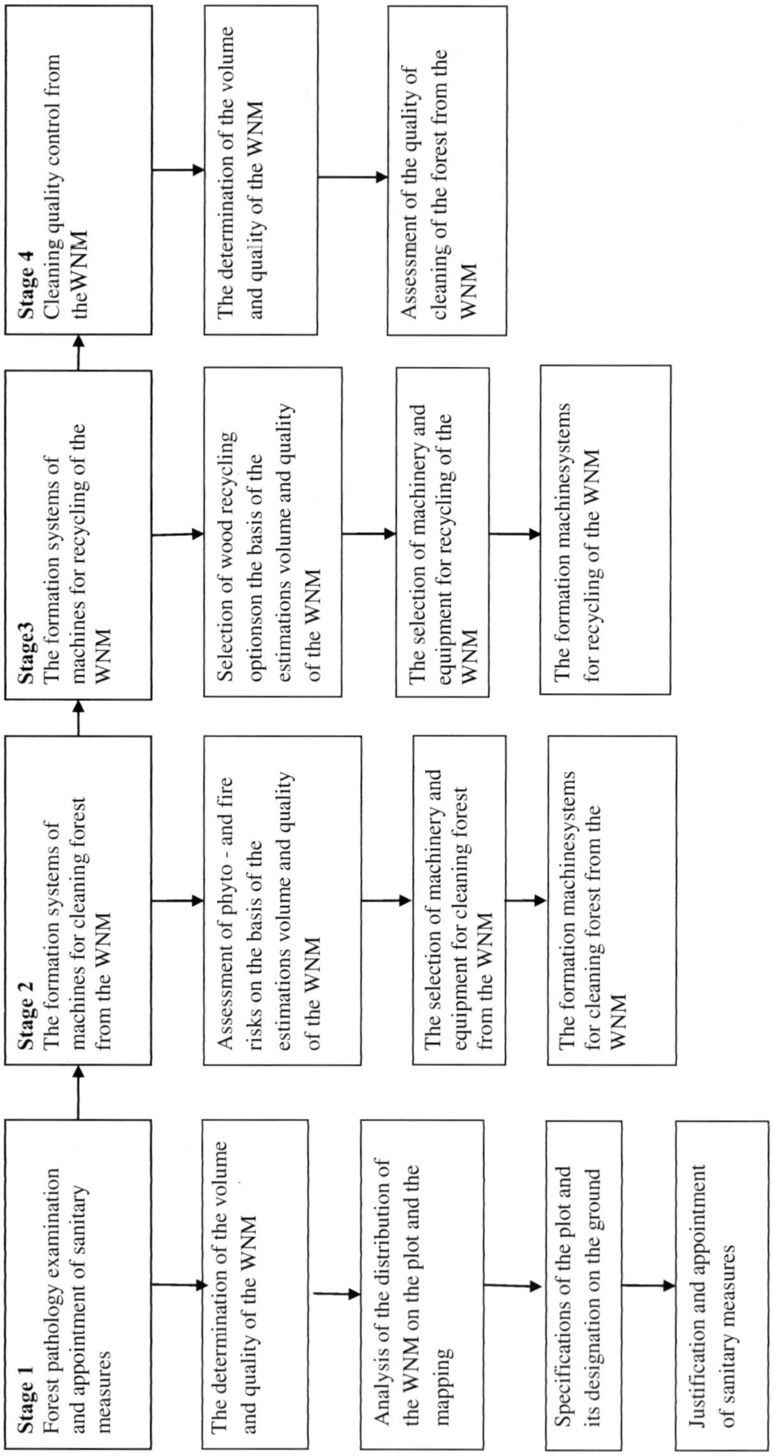

Figure 2. Block diagram of the solution of the tasks of cleaning forest from the WNM and recycling of the WNM.

IOP Conf. Series: Materials Science and Engineering **839** (2020) 012010 doi:10.1088/1757-899X/839/1/012010

In recent decades, technologies of low-volume logging operations based on specialized low-power machines and modular machines have been successfully recommended [2-13].

Specialized machines for low-volume logging are various mini-harvesters, mini-forwarders, mini-harvesters, etc. machines that perform the same technological operations as conventional logging machines, but have less power and some features depending on the purpose, for example, remote control.

In modular machines, the main process equipment is mounted or trailed, which is aggregated with the energy module. For example, a General-purpose wheeled tractor can act as an energy module. One energy module can be used for various logging and forestry operations. This significantly reduces economic risks and provides technological flexibility to a small enterprise. In the world practice, a lot of experience has been gained in the design of various trailed machines, from skidding trailers to harvester processors [12-13].

In the works of various scientists [2-13] shows the effectiveness of both small-size machines of the modular principle, and specialized machines for low-volume logging operations. All machines and equipment for cleaning forest from the WNM were divided into two groups:
1-on the basis of special forest machines and equipment;
2-on the basis of modular machines and equipment.

3. Description of the technological scheme of cleaning forest from the WNM and recycling of the WNM

General approaches were applied in the development of technology for cleaning forest from the WNM and recycling of the WNM for the conditions of the "Medyn forestry" enterprise.

Field surveys were established the WNM volume sand their quality. The volume of the WNM was determined by the line intersect method [1]. The main part of the WNM proved to be of poor quality and was classified as illiquid wood. It was decided to leave illiquid wood in the woodpile for further rotting in the forest. Part of the WNM of satisfactory quality was supposed to be sold to the population for firewood.

In the technological chain, the following types of the WNM were sequentially the subject of labor:
- unprocessed trees - processed trees (trunks) - skidding groups of trunks - stockpiles of trunks - crosscutted trunks (short logs) -stockpiles of short logs.

To perform most of the technological operations it was decided to use the existing forestry machines and equipment:
- processing the trees on the trunks for skidding - chain saws;
- skidding the groups of trunks to stockpiles - portable hand winch;
- crosscutting the trunks on the short logs and laying them in stockpiles - chain saws.

The process of cleaning up forest from the WNM is shown in Figure 3. The woodcutter with the chainsaw moves along the technological line and releases dangerous trees (dead, hanging, fallen trees, etc.). Woodcutter cuts the branches, getting the trunks. In addition, the woodcutter marks the trunks. Commercial and illiquid timber is marked with different colors.

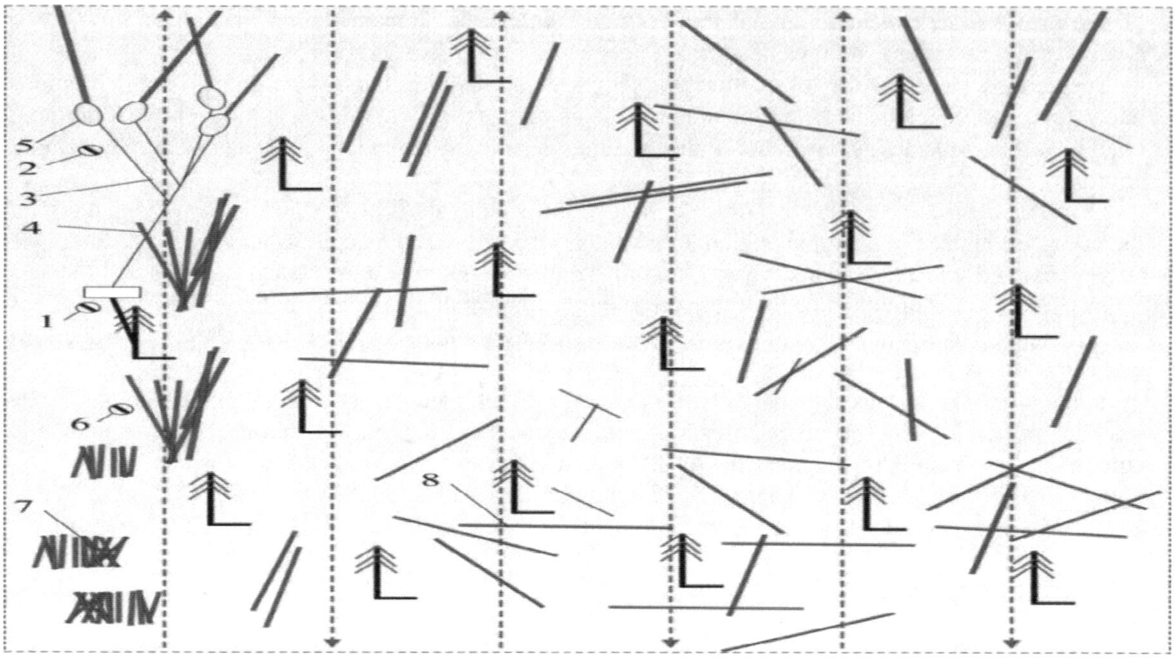

Figure 3. Cleaning forest from the WNM and recycling of the WNM: 1 – operator of the winch; 2 – woodcutter with the chainsaw on processing tree operation; 3 – traction rope; 4 – stockpiles of trunks; 5 – cones; 6 – woodcutter with the chainsaw on crosscutting operation; 7 – stockpiles of short logs.

The main objective of these activities is to prepare trees of the WNM for safe skidding in stockpiles of trunks. In the proposed technology, the trunks are crosscutted on the short logs and laid them in stockpiles of short logs. The skidding operation of the trunks is carried out as groups of several trunks. Equipment for trunks skidding includes:
- portable winch,
- plastic cones,
- durable polyester tapes and synthetic rope.

A portable winch is fixed using polyester straps to the tree. Trunks are choked through the cones in groups and skidded in stockpiles of trunks. The efficiency of the described technology was decided to be investigated on mathematical models.

Technological chains of forest cleaning from the WNM, according to the scheme in Figure 2, were divided into the steps of:
Phase 2. Woodcutter with the chainsaw + operator of portable winch.
Phase 3. Woodcutter with the chainsaw + stacker of shot logs into stockpiles.

This separation is due, in particular, to the fact that the two operations are performed separately. For example, the stockpiles of trunks formed by the winch can be in line and wait for processing them into shot logs for a long time without stopping the winch work. This article considers only the second stage of work (Figure 2), including the production chain: Woodcutter with the chainsaw + operator of portable winch.

4. Mathematical model of technological process

Productivity per shift of several chainsaws can be represented as the sum of the volumes prepared trunks for skidding per shift.

The time of the cycles and the productivity of the chainsaws per shift during the processing of the trees in the model are defined as the sum of all cycles and the volume of all processed trees:

$$\sum_{j=1}^{n}\cdot\sum_{i=1}^{N} t_{ji} = T\cdot\eta \Rightarrow Productivity\_per\_shift = \sum_{j=1}^{n}\cdot\sum_{i=1}^{N} q_{ji}, \tag{1}$$

where: t_{ji}– time of the i–th cycle of the j-thchainsaw, sec;

q_{ji} – volume of the j-th tree during the i-th work cycle, m^3;

N–number of complete cycles of one roller per shift;

T– duration of the shift, sec;

η - chainsaw utilization factor;

n– number of chainsaws.

The period of the cycle for tree processing is determined by the formula:

$$t = t_1 + t_2 + t_3 + t_4, \tag{2}$$

where:

t_1 – time for sawing twigs and the movement of the woodcutter from the butt to the top, sec;

t_2 – time to turn the trunk, sec;

t_3 – time for sawing supporting branches, sec;

t_4 – time for the transition of the woodcutter between the trees, sec.

Productivity per shift of the winch can be represented as skidded sum of the volumes of groups of trunks. The volume of the group of trunks is determined by the limit skidding volume of trunks. In the limiting case, the group may consist of a single trunk. The volume of the stockpiles of trunks is defined as the skidding sum of groups of trunks from one working station of the winch. This volume is determined by the covered winch area, which depends on the length of the traction rope:

$$Q_{group} = q\cdot\frac{\pi\cdot l_{rope}^2}{20000}, \tag{3}$$

where:

l_{rope}– length of the traction rope winch, m;

q– volume of the WNM per 1 ha, m^3/ ha.

The time of the cycles and the productivity per shift of the winch in the model is defined as the sum of all cycles and the volume of all skidded groups of trunks by the winch:

$$\sum_{i=1}^{N} t_i = T\cdot\eta \Rightarrow Productivity\_per\_shift = \sum_{i=1}^{N} Q_{i\_group}, \tag{4}$$

where:

t_i– time of the i–th cycle of the winch, sec;

Q_{i_group}– volume of the group of trunks during the i-th winch operation cycle, m^3;

N– number of complete cycles of operation per shift;

η– winchutilizationfactor.

The time duration of the i-th skidding cycle of one group of trunks:

$$T_{i\_group} = (t_1 + t_2 + t_3 + t_4)_i, \tag{5}$$

where:

t_1, t_2 – time of movement of the winch traction rope, respectively, without and with the load, c.

$$t_1 + t_2 = \frac{l}{V_1} + \frac{l}{V_2}, \tag{6}$$

where:

V_1, V_2 –speed of movement of the winch traction rope, respectively, without and with the

IOP Conf. Series: Materials Science and Engineering **839** (2020) 012010 doi:10.1088/1757-899X/839/1/012010

load, m/sec.

l– average skidding distance, m;

t_3– time for formation of the group of trunks for skidding, sec;

t_4– time to unload the group of trunks after skidding, sec.

With one working station,the winch is skiddingM_{group} groups of trunks withthe volume Q_{group} each. These groups form the skidding sum of groups of trunkswith the volume of Q_{pile}:

$$M_{group} = \frac{Q_{pile}}{Q_{group}}, \tag{7}$$

The time duration of thc j-th cyclc of the winch on the formation of the j-th group of trunks:

$$T_{j\_group} = \left(\sum_{i=1}^{M_{group}} T_{i\_group} \right)_j + t_{j\_winch}, \tag{8}$$

where:

t_{j_group}–time to relocate the winch after the formation of the j-thgroup of trunks, sec.

5. A conceptual model of the technological process

We will present the process as a Queuing system (QS).

Technology process can be presented in the form of a diagram in Figure 4. In terms of QS the service requests are trees– processed trees (trunks) group of trunks – stockpiles of trunks. Service devices are chainsaws on processing of the trees for skidding (K_{1i}) and a portable winch on skidding of groups of trunks to the stockpiles of trunks (K_2). With the beginning of the work, the system receives one demand, which generates n service requests, equal to the number of unprocessed trees in the working area of the winch. These are the service requests of the first level.

The service requests of the first level are served through channels K_{1n}. This is multi-channel service unit. It simulates the time of trees processing for skidding. The service time in the i-th channel K_1 is determined by the work of the chainsaw.

The accumulated service requests of the first level form the service request sof the second level – groups of trunks. The second level service request sare queued to the $Q_{group}$and served through the channel K_2. After the channel K_2, the demand of the second level are accumulated and form the service requests of the third level Q_{pile}, which simulate the stockpiles of trunks. Channel K_3 simulates the relocation of the winch to the new work station. After going through the service channel K_1, the service requests of the first level are placed in the queue and form the service requests of the second level Q_{group}.

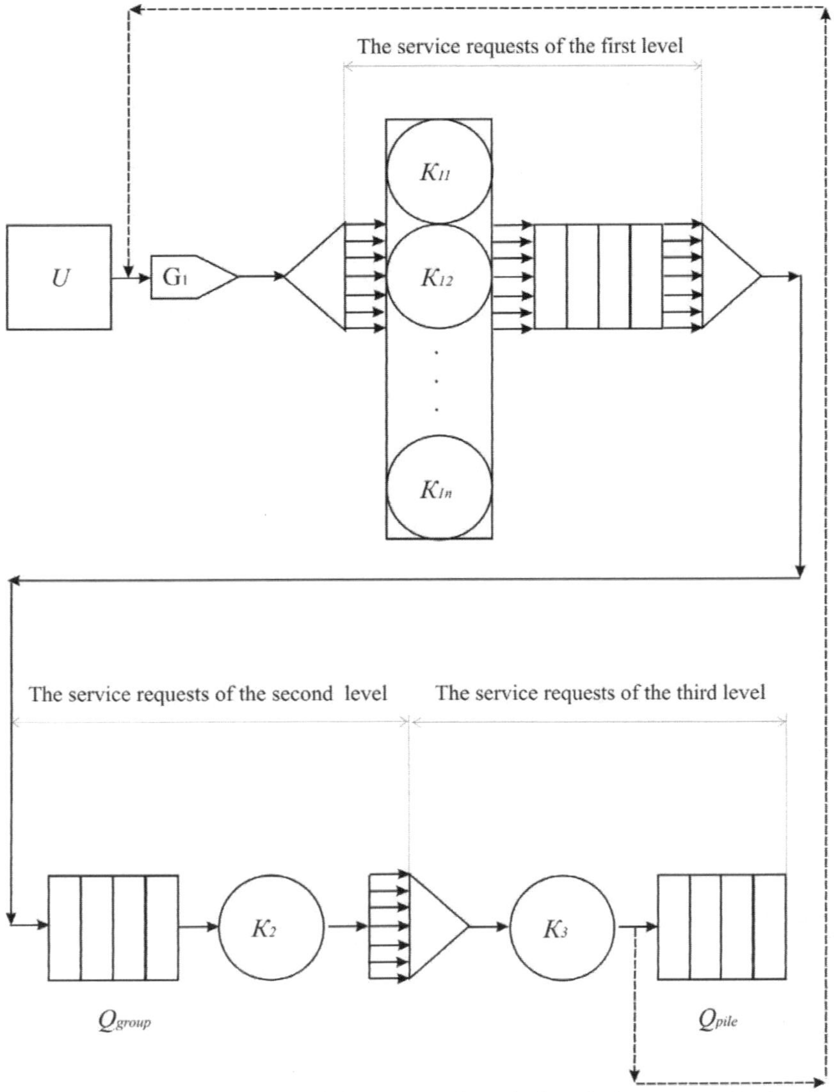

Figure 4. The scheme of the processing chain: chainsaws + winch.

Thus, in the model adopted by us, service requests are divided into three levels:
-service requests of the first level –unprocessed trees of the WNMand processed trees;
-service requests of the second level – skidding groups of trunks;
-service requests of the third level – stockpiles of trunks.

6. Planning of experiments and analysis of research results

The main objective of the experiments with the simulation model of the first technological stage was to determine the effect of the WNM volume on the forest plot on the process performance.

In the first series of experiments, the task was to establish the effect of the characteristics of the winch (the length of the winch traction rope, and hence the covered area, and the speed of the traction rope) on the performance of the technology.

The generalized results are given in Table 1-2, the experimental results were systematized and presented in the form of graphs in Figure 5-8.

Table 1. Results of experiments with simulation model 1 (1 chainsaw, 1 winch, volume of the WNM = 20 m³/ha).

Winch rope length, m	Winch rope speed, m/sec	The utilization of a winch	The utilization of a chainsaw
10	0,2	0,618	0,647
30	0,2	0,947	0,684
50	0,2	0,983	0,542
70	0,2	0,992	0,444
90	0,2	0,995	0,375
10	0,4	0,585	0,666
30	0,4	0,938	0,813
50	0,4	0,979	0,684
70	0,4	0,989	0,581
90	0,4	0,994	0,500

Table 2. Results of experiments with simulation model 1 (continuation) (1 chainsaw, 1 winch, volume of the WNM = 20 m³/ha).

Winch rope length, m	Winch rope speed, m/sec	The volume formed by the group of trunks, m³	Number of trees of the WNM on the forest plot	The volume of groups of trunksskidded per shift, m³/shift	The total numbers of groups of trunks skidded per shift, m³/shift
10	0,2	0,300	2	17,220	54,40
30	0,2	2,702	18	18,206	6,73
50	0,2	7,837	52	14,400	1,84
70	0,2	15,396	102	11,739	0,76
90	0,2	25,500	169	9,881	0,39
10	0,4	0,300	2	17,738	59,13
30	0,4	2,703	18	21,626	8,00
50	0,4	7,824	52	18,191	2,33
70	0,4	15,403	102	15,403	1,00
90	0,4	25,350	169	13,309	0,33

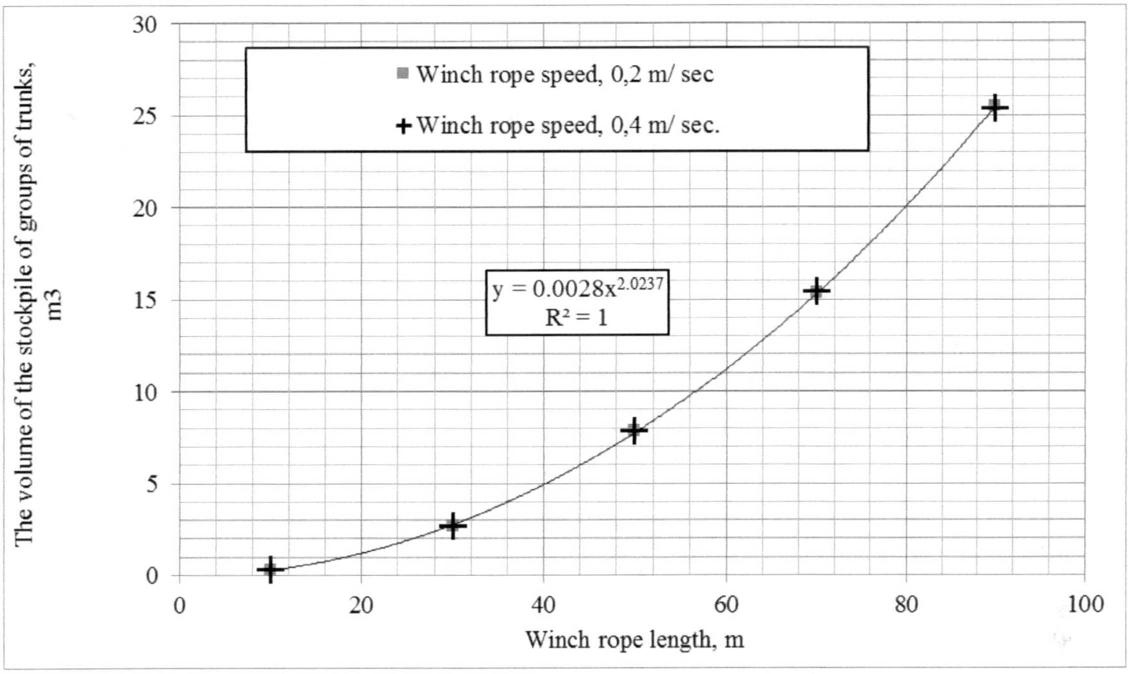

Figure 5. Dependence of the volume of the stockpile of groups of trunks formed on one working station of the winch, from the length of the traction rope winch (1 chainsaw, 1 winch, volume of the WNM = 20 m^3/ha).

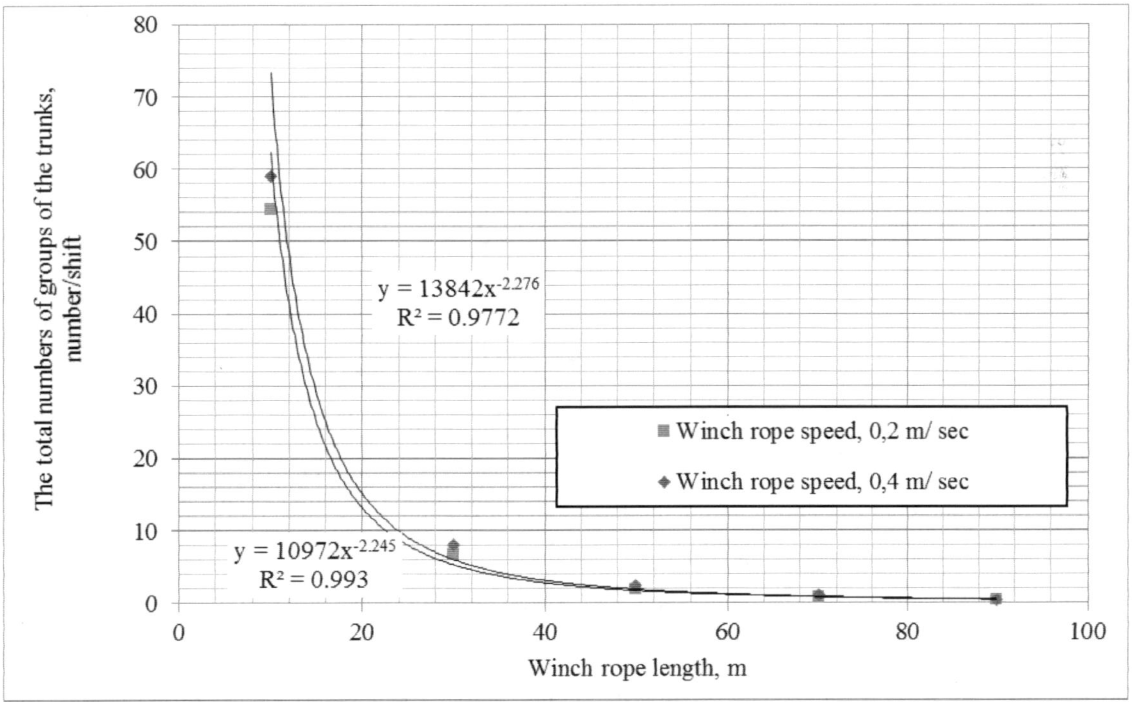

Figure 6. Dependence of the total numbers of groups of the trunks formed on one working station of the winch per shift, from the length of the traction rope winch (1 chainsaw, 1 winch, volume of the WNM = 20 m^3/ha).

Note that the speed of the winch rope does not affect the volume of the formed groups of trunks on one working station of the winch, which is understandable. From the chart in Figure 5 it can be seen that the volume of the formed stockpile of the trunks from one working station of the winch depends on the length of the winch traction rope. Note that with the increase of the length of the rope, the number of trunks in the working zone of the winch increases according to the power law. And the degree is close to 2. This was to be expected, since the area of the operation is proportional to the square of the length of the traction rope of the winch.

The graph in Figure 6 is very indicative. The number of groups of the trunks skidded per shift decreases sharply (according to the power law) with the increase of the length of the traction rope of the winch. This was to be expected, since the volume of the stockpile of the trunks and hence the time to form it, increases with the length of the winch rope (Figure 6).

The graph in Figure 7 is very interesting. (The dependence of the skidded volume of groups of the trunks from the length of the traction winch rope).

The curves in the graph Figure 7 have maximums that is, at some length of the winch traction rope, the system achieves maximum performance. In the case of the WNM volume of 20 m^3/ha, the maximum capacity of the system will be achieved at 30-meter rope length.

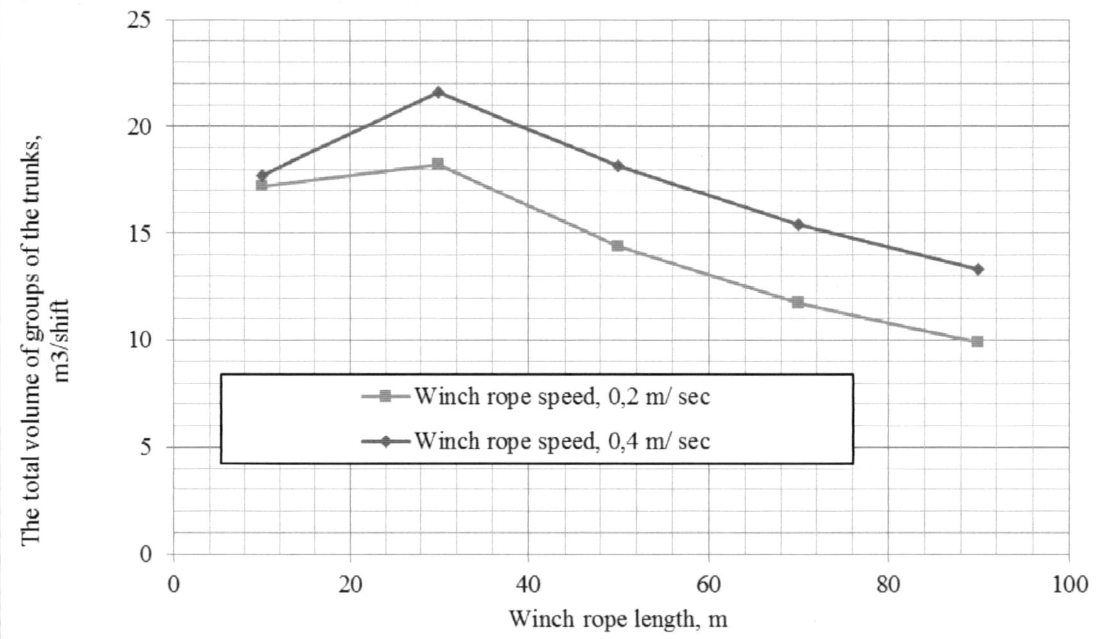

Figure 7. Dependence of the total volume of groups of the trunks formed per shift from the length of the winch traction rope (1 chainsaw, 1 winch, the volume of the WNM = 20 m^3/ha).

On Figure 8 the graph of the dependence of coefficient of utilization of the winch and the chainsaw from the length of the traction rope of the winch is given. Analysis of curves of the chart Figure 8 shows that the maximum values of the coefficients of utilization of the winch and the chainsaw are reached at 30-meter length of the traction rope. And this occurs both at the speed of the winch traction rope equal to 0,2 m/sec, and at 0,4 m /sec.

Note that the coefficient of utilization of the winch (Figure 8) from the length of the traction rope winch grows rapidly with increasing traction rope length and at 40 m becomes close to 1.

In order to establish the effect on the coefficients of utilization of the winch and the chainsaw from the volume of the WNM, an additional series of experiments was carried out.

IOP Conf. Series: Materials Science and Engineering **839** (2020) 012010 doi:10.1088/1757-899X/839/1/012010

The planning matrix of the experiments is given in Table 3. The results of the experiments were systematized and presented in the form of a graph in Figure 9.

From the chart in Figure 9 follows that with the increase of the volume of the WNM, the coefficient of utilization of the winch increases, and the coefficient of utilization of the chainsaw decreases. However, these changes are insignificant (no more than 3,5%) and may not be taken into account in practice. Therefore, for the conditions adopted in the experiments, the length of the traction rope equal to 30 m can be considered optimal for portable hand winches.

Table 3. Planning matrix of experiments with simulation model 1 (1 chainsaw, 1 winch, traction rope length = 30m, rope speed = 0,2 m/sec).

Winch rope length, m	Winch rope speed, m/ sec	Volume of the WNM, m^3/ha
30	0,2	20
30	0,2	40
30	0,2	60

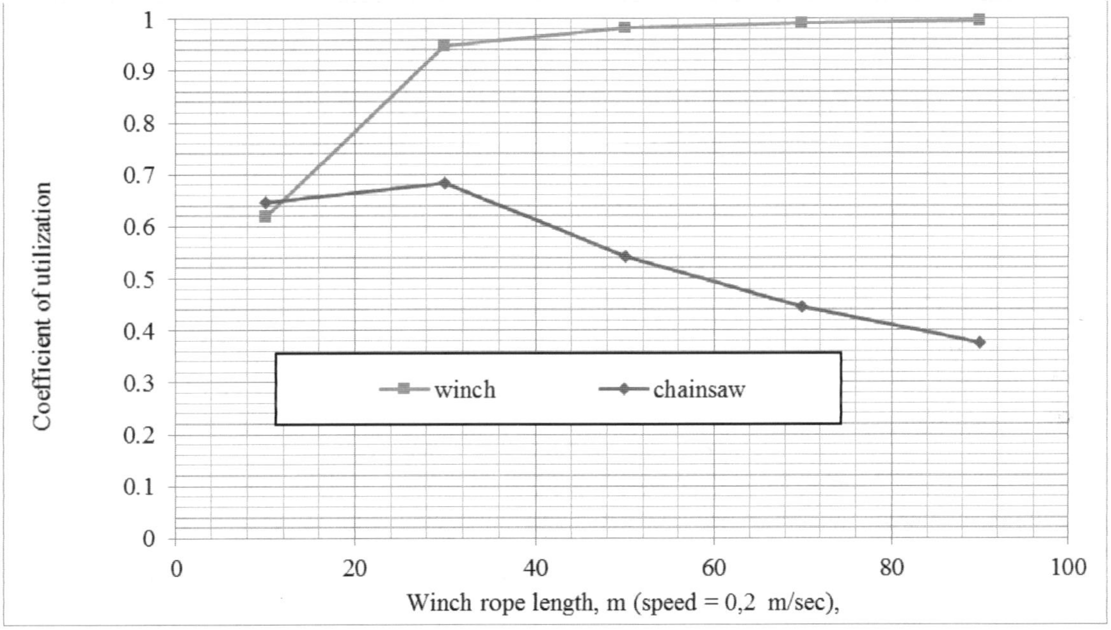

Figure 8. Dependence of the coefficients of utilization of the winch and the chainsaw from the length of the traction rope of the winch (1 chainsaw, 1 winch, the volume of the WNM = 20 m^3/ha).

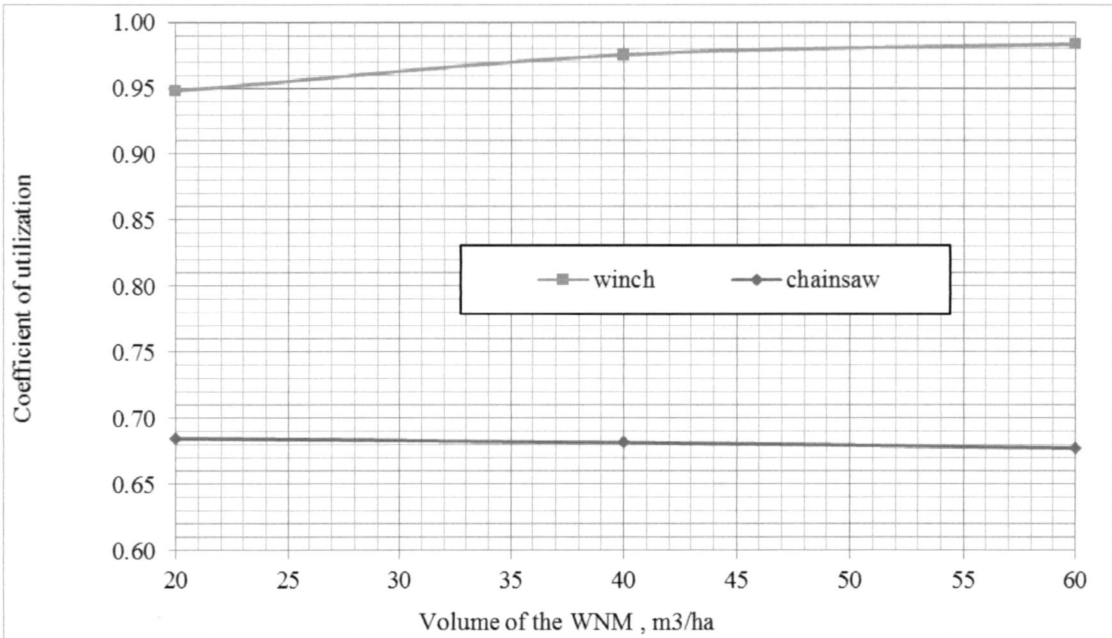

Figure 9. Dependence of the coefficients of utilization of the winch and the chainsaw from the volume of the WNM (1 chainsaw, 1 winch, rope length = 30m, traction rope speed = 0,2 m/sec).

7. Conclusion and practical recommendations

The simulation experiments that were carried out with the dead wood or wood natural mortality (WNM) made it possible to draw the following conclusions:

1. The technology cleaning and utilization of the DEO is proposed, which consists of four stages, including quality control of sanitary measures (Figure 2).

2. In the framework of the proposed multi-stage technology, the article considers the second stage of the DEO cleaning (Figure 2), including the process chain: chainsaw + portable winch (Figure 2).

3. A mathematical and conceptual model of cleaning the DEO was developed. The model considers a winch with a rope length from 10 to 90 m and DEO volume on a site from 20 to 60 m^3/ha.

4. It is established that the volume of the DEO from one winch parking depends on the length of the winch traction rope. As the length of the rope increases, the number of the DEO that contains inside the winch's operating zone increases by a power law.

For practice, we can conclude that the length of the winch rope should be chosen taking into account the volume of DEO on the site (Figure 5).

5. Maximum performance is achieved with a certain length of the winch traction rope. For example, in the case of the DEO volume of 20 m^3/ha, the maximum productivity of the system will be achieved with a 30-meter rope length (Figure 7).

In practice, this means that the length of the winch rope should be chosen based on the volume of the DEO on the site and the winch power.

6. With the increase of the DEO volume on the site, the winch utilization rate increases, and the chainsaw utilization rate of the chainsaw falls. However, these changes are insignificant (no more than 3.5%) and may not be taken into account in practice (Figure 9).

In practice, for the conditions adopted in the experiments, the length of the traction rope equal to 30 m can be considered optimal for portable hand winches.

8. References

[1] Karpachev S P, Zaprudnov V I, Bykovskiy M A and Scherbakov E N 2017 Quantitative estimation of logging residues by line-intersect method *Croat. J. For. Eng.* **38**(1) 33-452

[2] Karpachev S P and Diev R I 2017 Modelirovanie tekhnologicheskogo protsessa uborki zakhlamlennosti lesa portativnoy lebedkoy [Modeling of technological processes of cleaning litter the forest with portable winch] *Lesnoy vestnik / Forestry Bulletin* **21** 10-14
[3] Karpachev S P and Diev R I 2017 Process simulation of wood mortality cleaning for needs of bioenergetics *Machinery and equipment for the village* **6**(240) 46-48
[4] KarpachevS P and Diev R I 2017 Wood harvesting technology development of natural losses for bio-energy needs *Machinery and equipment for the village* **3**(237) 41-43
[5] Karpachev S P, Zaprudnov V I, Shmyrev V I, Shmyrev D V, Kamusin A A and Red'kin A K 2017 Technological process simulation for development of forest bio-resources using soft containers at logging sites *Machinery and equipment for the village* **2**(236) 45-48
[6] Di Fulvio, Bergström D, Kons K and Nordfjell T 2012 Productivity and profitabilityof forest machines in the harvesting of normal and overgrown willow plantations *Croat. J. For. Eng.* **33**(1) 25–37
[7] Eriksson M and Lindroos O 2014 Productivity of harvesters and forwarders in CTL operations in northern Sweden based on large follow-up datasets *Int. J. Forest Eng.* **25**(3) 179-200
[8] Nilsson B 2016 Extraction of logging residues for bioenergy-effects of operational methods on fuel quality and biomass losses in the forest *Linnaeus University Dissertations* No 270 200 p
[9] Mihelič M, Spinelli R and Poje A 2018 Production of Wood Chips from Logging Residue under Space-Constrained Conditions *Croatian J. Forest Eng.* **39** 223
[10] Sukhanov Yu, Yu Gerasimov and Selivyorstov A 2013 Efficiency of Forest Chip Supply Systems in Northwest Russia *Adv. Mat. Res.* **740** 799
[11] Laitila J, Heikkilä J and Anttila P 2010 Harvesting alternatives, accumulation and procurement cost of small-diameter thinning wood for fuel in Central Finland *Silva Fennica* **44** 465
[12] Schweier J, Spinelli R, Magagnotti N and Becker G 2015 Mechanized coppice harvesting with new smallscale fellerbunchers: Results from harvesting trials with newly manufactured felling heads in Italy *Biomass Bioenerg.* **72** 85
[13] Chakroun M, Bouvet A, Ruch P and Montagny X 2016 Performance of two shear heads for harvesting biomass in hardwood stands in France *Biomass Bioenerg.* **91** 227

IOP Conf. Series: Materials Science and Engineering **839** (2020) 012011 doi:10.1088/1757-899X/839/1/012011

Compressive strength and truck run over ability of plastic/sand paving block composites

I Mashudi[1], N P G Suardana[1,*], I N Arya Thanaya[2], I W Bandem Adnyana[1] and C I P K Kencanawati[1]

[1] Mechanical Engineering Department, Udayana University, Bali, Indonesia
[2] Civil Engineering Department, Udayana University, Bali, Indonesia
[*] Corresponding author: npgsuardana@unud.ac.id

Abstract. Plastic waste has a negative effect on the environment since it could not be decomposed easily, and it could decrease soil fertility. One way to decrease plastic waste is by mixing it with sand to make paving blocks, because the characteristic of plastic is elastic and flexible, so it is expected that the material will be ductile and strong. Material of this study used LDPE (Low Density Polyethylene) plastic which was chopped and mixed with 6 mesh sand with variation of 1 kg (plastic): 5 kg (sand) and 1 kg (plastic): 7 kg (sand) (1:5 and 1:7). The sand was inserted into mixing machine with 200 °C temperature for 5-10 minutes, after that the chopped plastic was inserted and mixed for 30-35 minutes. Molding process was done by pressure tool. After that testing was done by using compressive machine and run over vehicle. The research result showed that the lowest damage occurred on plastic and sand composition of 1:5 as much as 0.113% and had compressive strength 16.667 MPa, and the highest damage occurred on plastic and sand composition of 1:7 as much 0.267% and had compressive strength 12.963 MPa.

Key words: paving block, LDPE plastic, run over vehicle testing, compressive strength

1. Introduction

Waste problem has not been solved comprehensively in Indonesia until now, by the increase of population, the volume of waste is increasing as well as the consequence of human activity. Most of human activities produce organic waste as much as 60-70% and non-organic waste of 30-40%, in which 14% from the non-organic wastes are plastic wastes [1]. At the first time, plastic was invented in 1907, since then the use of plastic and plastic-based material are increasing. Plastic could not be separated from industrial and consumerism development. Many industries in this world use plastic to pack their products. Instant food industry for instance choose to use aluminum foil-coated plastic or multilayer plastic for their packages because it is considered safe and keep the product consumable [2]. At the same time, the industrial products increase, the human needs become more complex, the people keep shopping, and needs bag for carrying their goods. Plastic bag became the answer for this issue because it is inexpensive, strong, light, not rusty, thermoplastic, could be labeled with many creations, and it could be an effective branding media.

In other side, the plastic wastes bring negative effect on the environment and human health. It has negative effect on environment because it takes longer time to be decomposed and it could decrease soil fertility. Littering plastic waste will also block drainage, sewer and river so there will be flood. The burnt plastic waste will produce Volatile Organic Compounds (VOCs), smoke (particulate), heavy metal bonded particles, polycyclic aromatic hydrocarbons (PAH), polychlorinated dibenzofurans (PCDFs) and dioxin, in which those substances are dangerous for human and environment [3]. The increasing of plastic users will trigger many problems for environment. The common way of reducing plastic waste used by people is 3R (Reuse, Reduce, and Recycle).

Other alternative to solve plastic waste problem is mixing plastic waste with concrete, paving block and asphalt because the characteristic of plastic is elastic and flexible, so it is expected that the material will be stronger. Nowadays many people use paving block as building construction, especially for road pavement in the neighborhood, houses, parking lot and other places. Paving block is widely used because it could withstand load within certain limits and easily installed. Besides that, paving block is better than other pavements concerning the economic value of maintenance, exterior artistic aspect of the building, do not need heavy tool, and it can be massively produced as well as concerning the environmental sustainability it could preserve water. The use of cement as the main material of paving block is relatively expensive. Therefore, this study used plastic bag waste as the replacement of cement, which is expected that low density polyethylene (LDPE) plastic bag waste could produce paving block that is able to withstand loads of public vehicle. Many researchers studied the mixing bitumen with plastic, but in the present work plastic bag waste and sand pavement are studied with different compositions.

2. Research Method

This study used LDPE plastics which were washed and dried under sunshine for 24 hours, and then they were chopped by using plastic chopper. Besides plastic waste, other main material of this study was sand. It was a river sand that had been cleaned and sifted using a 6-mesh sieve. The sifted sand was put into mixer machine and was heated in 200 °C for 5-10 minutes, after that the chopped plastic was put into the mixer and mixed for 30-35 minutes with composition comparison of plastic: sand (1 kg:5 kg and 1 kg: 7 kg).

The mixture of plastic and sand was molded by using a beam-shaped iron plate with a size of 20 × 20 × 6 cm in accordance with the shape and size of the paving [4] then it was pressed by using hydraulic jack and thick plate as a cover to avoid voids. After cooling, the specimen was pulled out from the mold and keep on the room temperature minimum for 24 hours before testing and labelled based on the molded mixture.

For compressive test, paving block was cut into cubes sized 6 × 6 × 6 cm in accordance with the ASTM D695 [5] and compressive testing was done by using mortar compression machine UT200/50EL. For run over testing, the paving block was installed on field in accordance with standard of paving installation [6] and then it was run over using a truck without a load with variations in the run of 10, 50,100, 200, 400 times. For every run variation, the paving was measured to know the damage and the damage mean score was calculated in form of percentage.

3. Results and Discussion

3.1. Compressive Test

Compressive test was done by using mortar compression machine UT200/50EL to know the compressive strength of the paving block. The following (Table 1) is the table of paving compressive strength calculation:

Table 1. Compressive strength of plastic paving and concrete paving.

Comparison	Sample	Compressive load (KN)	Area (cm^2)	Compressive strength (MPa)	Mean score (MPa)
Plastic: sand (1:5)	Specimen 1	60	36	16.667	16.667
	Specimen 2	65	36	18.056	
	Specimen 3	55	36	15.278	
Plastic : sand (1:7)	Specimen 1	55	36	15.278	12.963
	Specimen 2	40	36	11.111	
	Specimen 3	45	36	12.500	
Cement : sand (1:5)	Specimen 1	95	36	26.389	23.148
	Specimen 2	70	36	19.444	
	Specimen 3	85	36	23.611	
Cement : sand (1:7)	Specimen 1	55	36	15.278	17.129
	Specimen 2	80	36	22.222	
	Specimen 3	50	36	13.889	

Then, the data were plotted into graphic to describe the mean score of the compressive strength comparison of plastic paving and concrete paving as follows:

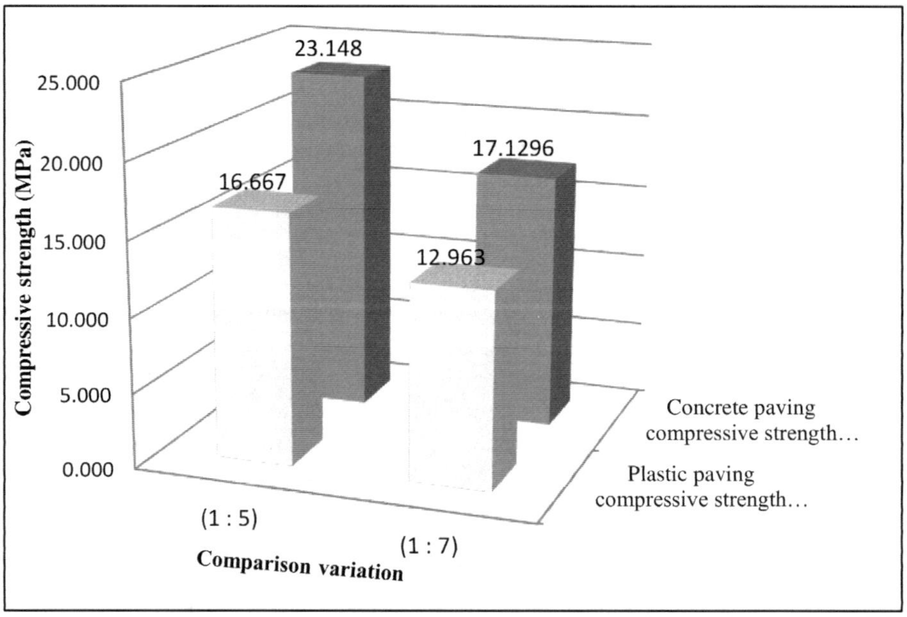

Figure 1. Graphic of compressive strength comparison of plastic paving and concrete paving.

Based on the above mean score of compressive strength from every samples as shown in Table 1 and Figure 1, the difference of plastic and sand or cement and sand mixture affected the paving compressive strength. The highest compressive strength occurred in cement paving with comparison of cement and sand 1:5, as much as 23.148 MPa. Meanwhile, the lowest compressive strength occurred in plastic paving with comparison of plastic and sand 1:7, with compressive strength of 12.963 MPa. It is in line with research conducted by Frigione [7] who replaced the sand with polyethene waste that the compressive and tensile strength of the plastic concrete was 0.4 – 1.9% which was lower than control but with slightly higher ductility.

IOP Conf. Series: Materials Science and Engineering **839** (2020) 012011 doi:10.1088/1757-899X/839/1/012011

Figure 1 shows that compressive strength of plastic paving is lower than cement paving, but plastic paving specimens were not destroyed after compressive testing as shown in Figure 2 and Figure 3, while concrete paving specimens were destroyed as shown in Figure 4 and Figure 5. It means that plastic paving block was more persistent than cement paving block.

Based on National Standardization Agency of Indonesia(*BSN*) [8] compressive strength qualities for paving block standard B (for parking lot) and standard C (for pedestrian) are 17.0 MPa and 12.5 MPa, respectively. Therefore, plastic paving with mixture 1:5 and compressive strength 16.667 MPa almost meet the BSN standard and plastic paving with mixture 1:7 and compressive strength 12.963 MPa has already met BSN standard. In Agyeman et.al [9] research, composite paving blocks less in plastic (LP) has compressive strength 7.31 N/mm^2 (MPa) and composite paving blocks high in plastic (HP) has compressive strength 8.53 N/mm^2 (MPa). So, plastic paving with mixture 1:5 and 1:7 have higher compressive strength.

Figure 2. Plastic paving 1:5.

Figure 3. Plastic paving 1:7.

Figure 4. Concrete paving 1:5.

Figure 5. Concrete paving 1:7.

3.2. Run Over Vehicle Test

This test was conducted by using double ankle truck Hino Dutro Dump 130 HD without loading with weight 2,355 kg and standard tire pressure of 100 psi. The truck run over the specimen by going forward and backward, on every variation of run, the specimens were measured to know the damage.

Mean score of damage was changed into damage percentage and plotted into graphic of the relation between number of run and the damage occurred as follows:

IOP Conf. Series: Materials Science and Engineering **839** (2020) 012011 doi:10.1088/1757-899X/839/1/012011

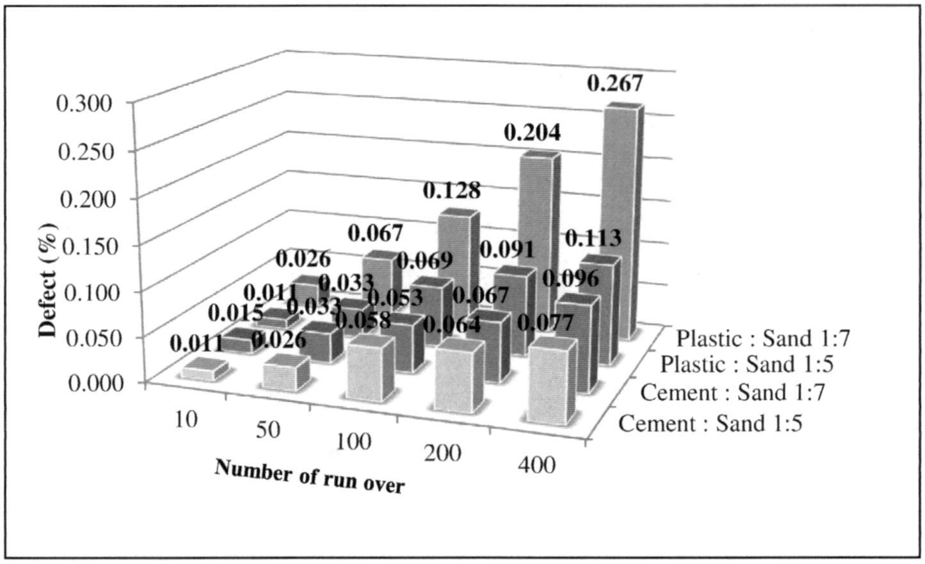

Figure 6. Graphic of run over damage comparison between plastic paving and concrete paving with comparison variation 1:5 and 1:7.

The data of run over test which was done with mixture variation of LDPE plastic and sand, produces data of damage percentage as shown in Figure 6. It shows that the difference of plastic and sand mixture affects the number of damage percentages. Based on Figure 6, it can be seen that the lowest damage occurred in concrete paving with composition of cement and sand 1:5 as much as 0.073%, yet in the testing the concrete paving was damaged in two parts on run 200th while the plastic paving with composition of plastic and sand 1:5 has damage average 0.113%.The highest damage occurred in plastic paving with composition of plastic and sand 1:7 as much as 0.267%, but there is no broken. From all data, there is no big difference of damage percentage or not too significant for unbroken specimens. Kosmatka et al. [10]_argued that concrete resistance upon impact and abrasion related to compressive strength and aggregate type. From this argument, it can be proven that plastic paving mixture 1:5 which has compressive strength 16.667 MPa has lower damage than plastic paving 1:7 which has compressive strength 12.963 MPa.

3.3. The result of material characteristic using SEM (Scanning Electron Microscope) Type Phenom G2 Pro.

Both results of testing were supported by SEM photo of plastic paving with mixture comparison variation 1:5 and 1:7 as the following Figure 7 and Figure 8.

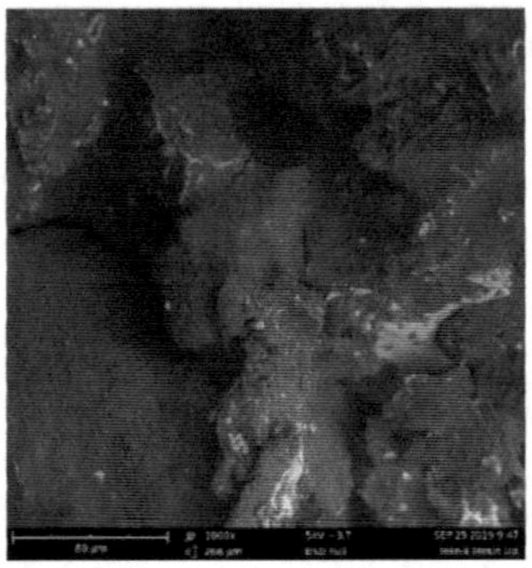

Figure 7. SEM of plastic paving 1:5.

Figure 8. SEM of plastic paving 1:7.

In which plastic paving with composition comparison of plastic and sand 1:5 has lower damage compared to plastic paving which has comparison plastic and sand 1:7. It is because on plastic paving 1:5 almost all of the sands were bonded (as shown in photo SEM Figure 7) so the strength to hold the load is better than plastic paving 1:7, in which paving with comparison 1:7 has more cavities caused by less of matrix to bond all surfaces of sand as shown by photo SEM Figure 8.

4. Conclusion

Based on the finding and discussion, some conclusions can be drawn as follows:

1. The higher mixture composition of plastic and sand, the lower compressive strength produced. The highest compressive strength of plastic paving was gained from mixture of plastic and sand 1:5 as much as 16.667 MPa, while the lowest compressive strength was gained in comparison 1:7 as much as 12.963 MPa.
2. The higher mixture composition of plastic and sand, the higher damaged occurs. The highest damage was gained from mixture of plastic and sand 1:7 as much as 0.267%, while the lowest damage was gained from comparison 1:5 as much as 0.113%.
3. Plastic-sand paving is lesser compressive strength but more ductile than cement-sand paving.

5. References

[1] Purwaningrum P 2016 Upaya mengurangi timbulan sampah plastik di lingkungan *J. Teknik Lingkungan* **8** 2

[2] Putra H P and Yuriandala Y 2010 Studi pemanfaatan sampah plastik menjadi produk dan jasa kreatif *J. Sains Dan Teknik Lingkungan* **2** 1

[3] Valavanidis A, Iliopoulos N, Gotsis G and Fiotakis 2008 Persistent free radicals, heavy metals and PAHs generated in particulate soot emissions and residue ash from controlled combustion of common types of plastic *J. of Hazardous Materials* **156** 277

[4] SNI 03-0691-1996 Standar Nasional Indonesia 03-0691-1996 tentang Bata Beton (Paving Block). *Badan Standardisasi Nasional* (1996)

[5] ASTM D695-15 Standard test method for compressive properties of rigid plastics, ASTM International, West Conshohocken, PA, 2015

[6] SNI 03-2403-1991 Standar Nasional Indonesia 03-2403-1991, tata cara pemasanganblok beton terkunci untuk permukaan jalan. *Badan Standardisasi Nasional* (1991)

[7] Frigione M 2010 Recycling of PET bottles as fine aggregate in concrete *J. Waste Management* **30** 1101

[8] Agyeman S, Obeng-Ahenkora N K, Assiamah S and Twumasi G 2019 Exploiting recycled plastic waste as an alternative binder for paving blocks production *J. of Case Studies in Construction Materials* **11** e00246

[9] Kosmatka S H, Kerkhoff B and Panarese W C 2002 Design and control of concrete mixtures 14th edition *Portland Cement Association* (USA: Skokie, Illinois) p 358

Acknowledgement

The gratitude is delivered to Directorate of Research and Community Services, directorate general for strengthening of research and development, Ministry of Research, Technology and Higher Education, based on Research Contract No: 492.65/UN14.4.A/LT/2019, date 11 March 2019.

IOP Conf. Series: Materials Science and Engineering **839** (2020) 012012 doi:10.1088/1757-899X/839/1/012012

The optimisation of fabrication process for various chemically treated kenaf fibres in epoxy matrix composites

Ghaztar M M M[1], Ibrahim N N I N[1, 2], Zakaria S[3] and Romli A Z[1, 2]

[1] Faculty of Applied Sciences, Universiti Teknologi MARA, 40450 Shah Alam, Selangor, Malaysia
[2] Centre of Polymer Composite Research and Technology, Institute of Science, Universiti Teknologi MARA, 40450 Shah Alam, Selangor, Malaysia
[3] School of Applied Physics, Faculty of Science and Technology, Universiti Kebangsaan Malaysia, 43600 UKM Bangi, Selangor, Malaysia

Abstract. The fabrication process is crucial since it determines the overall performance of the bio-composites. This study draws from the drawbacks of the natural fibre reinforced composites that are high voids content, low in surface interaction as well as low-stress distribution property especially using a simple mould-press process. The study explores the correlation between various fabrication processes of thermoset matrix towards the composites density, voids content, tensile and morphological properties. The study was carried out by mould pressed only 10 percent (%) filled composites by varying its curing temperatures, moulds type, fibres size and fibres chemical treatments afflicted. The results showed that 50 °C of curing temperature, in plunger mould-type, using 0.25 mm mesh size fibre are the optimised parameters. The coated 1:6 composite showing the optimum properties since it improves 6.5% density, 99.95% voids content, 38.1% water uptake and 33.2-50.6% of tensile properties. The suitable processing parameters and high coating penetration might contribute to the total improvement of the coated fibre composites properties. Since it influences the fibres compact-ability that leads to low voids formation as analysed on its fractured surfaces. In summary, a better understanding of the materials characteristic and fabrication processes are needed to obtain the optimum bio-composites performance.

1. Introduction

The prevalent usage of natural fibres as a reinforcing filler in various applications nowadays become a trend for the industrialist to introduce a 'green' element in their products. The benefits that are easily in availability and low tooling cost not just prolong to the manufacturer but, to the human and mother-nature as well since it less toxic and easily to decomposed [1]. But, the question remains on the low in properties of natural-based products since the fibres naturally have variability in physical and chemical aspects that affect its mechanical and morphological properties.

Therefore, the natural-based material and processing must be optimised to get an improved property to replace the available man-made materials. Recently, the biocomposites fabricator tend to manipulate the various processing parameters to achieve the said goal. However, by far, less research conducted on the combination of the parameters that referred to curing temperatures, mixing method, fibre size, type of mould and fibre surface treatment by considering the nature of the matrix and the fibre used. In this study, the epoxy thermosetting resin is used as the matrix of the composite since it recognised able to reduce its viscosity before its gel time.

This study also explores the effect of the heated mixing process between fibre and epoxy to its surface interaction. Since, in theory, the process able to make the matrix to become lower in viscosity and the important aspect, it enables to efficiently wet-out the surface of the natural fibre that naturally incompatible. The importance of the viscosity thinning is, it will improve the surface contact to make the matrix penetrated onto the delicate surfaces of the fibre. An improved matrices surface contact is crucial since it will determine the effectiveness of the stress transfer that cannot be achieved by a composite system that has low surface interaction.

It is also expected that, when the surface interaction is high, the fibre compaction inside the composite system also improved. Therefore, this study focuses on all the processing parameters that cover on material and composite fabrication aspects and only for thermoset matrix in the simple mould-press process. In an effort to optimise the fibre-filled biocomposites to perform at its best properties to replace the costly and harmful synthetic materials.

2. Methodology

2.1. Materials

The raw kenaf natural fibre was obtained by the Kenaf and Tobacco Board. The received fibre was in the continuous form (more than 30 cm in length). The epoxy resin (CP362A) and hardener (CP362B) were manufactured by Oriental Option Sdn. Bhd. The physical, mechanical and chemical characteristics of kenaf fibre and epoxy resin and hardener are tabulated in table 1 and 2 respectively. Acetone was supplied by Classic Chemicals Sdn. Bhd. with 95% purity and 58.08 g/mol of the molecular weight.

Table 1. The physical and mechanical properties of Kenaf bast fibres.

	Density (g/cm^3 fibres)		Lengths (mm)	Diameter (μm)		Length/ diameter		Tensile Properties	
Fibrous Material	Fibres	Bundle	Range	Average	Range	Average	Ratio	Average and Spec. Strength (MPa)	Average and Spec. Modulus (GPa)
Kenaf (bast)	-	1.2	1.4-11	2.6	12-36	21	124	295-930 & 246-993	22-60 & 18-50

Source: Physical and mechanical properties of commercially important lignocellulosic fibre and of E-glass fibre [2]

Table 2. The physical characteristics of CP 362 epoxy resin.

Part	Chemical Type	Viscosity (cps)	Colour	Gel Time	Post Cure	Final Viscosity (cps)
Epoxy (CP 635A)	Epoxy DGEBA	13 000	Transparent	35 Minutes (25°C)	9.5 Hours (25°C)	8500
Hardener (CP 362B)	Modified Polyamine	400	Transparent			

Source: All Purpose Epoxy [3]

2.2. Fibre Treatment Process

Before the coating treatment, the Kenaf fibres were combed prior conditioned at 24 °C for 48 hours. The coating solutions were prepared by diluting the epoxy solution with acetone at 1:4, 1:5 and 1:6 w/w epoxy to acetone ratios. The fibres were immersed instantly after prepared using 3:7 fibre to coating ratio in room temperature. The processes were conducted in a controlled condition to prevent evaporation of acetone. After 5 minutes of soaking process, the fibres were tossed and individually separated to promote a single fibre coating during the curing process at 80 °C for 24 hours in an air circulated oven.

The untreated and sodium hydroxide (NaOH)/ aminosilane (APS) treated fibre batches become reference 1 and 2 samples respectively. Prior treatments, the fibres were combed and cleansed using water and dried. For the first treatment, the fibres were immersed in 6 w/v % of NaOH-water solution with 7:3 immersion ratio at 25 °C. The fibres were washed and dried at 70 °C at 96 hours. The fibres were treated again with 1 (v/v %) of APS, 95 (v/v %) of ethanol, 4 (%) water for 30 minutes and air-dried for 1 hour.

2.3. Pulverisation Process

All fibres were cut using a pulveriser machine (Fritsch Power Cutting Mill PULVERISETTE 15) passing thru 5.0 mm, 1.0 mm, 0.5 mm until 0.25 mm pulverise mesh sizes aperture in consecutively. The fibre was added gradually and collected in the receiver bowl. Only, the 5.0 mm and 0.25 mm sizes of pulverised fibres were used in this study.

2.4. Fabrication Process

The details of the process flow of tests and analyses were illustrated in figure 1 and 2. The summary of the analysis parameters was tabulated in table 3. All the composite samples were pressed at 1200 psi using HP-25T press machine for 2 hours and post-cured at 50 °C for 24 hours using an air circulated oven.

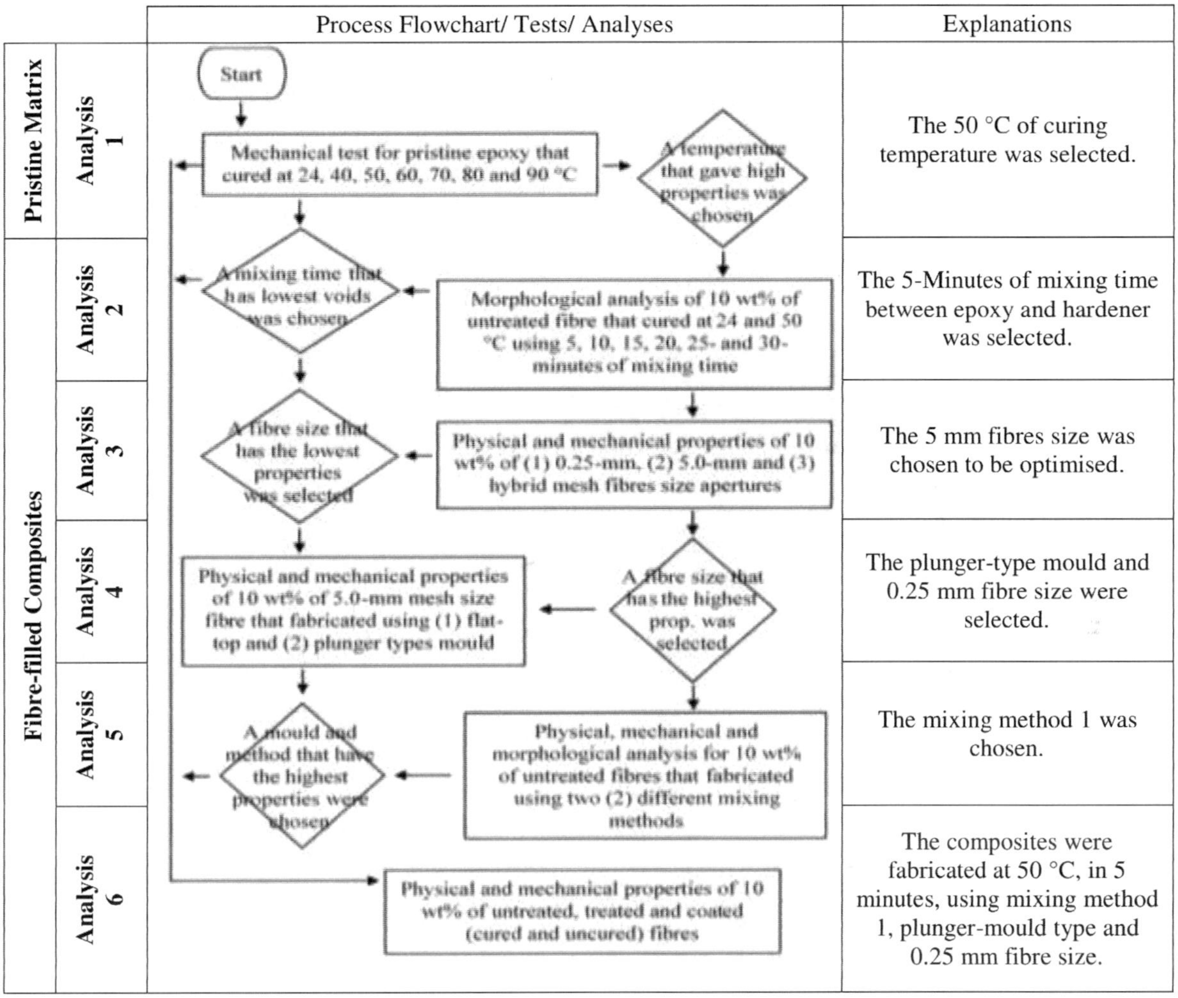

Figure 1. Process flow chart for test and analysis.

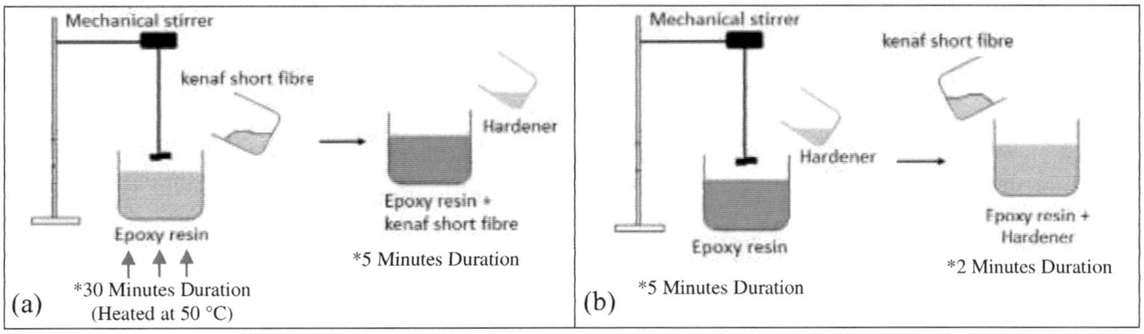

Figure 2. Illustrations of mixing (a) method 1 and (b) method 2.

Table 3. The overall fabrication parameters in all analyses.

Analysis	Fibre Loading (wt%)	Composites Curing Temperatures (°C)	Matrices mixing time (mins)	Fibre Sizes (mm)	Mould Types	Mixing Methods	Fibre Treatments
1	Nil.	24 to 90	5	Nil.	Flat	Nil.	Nil.
2	10	24 and 50	5 to 30	0.25	Flat	2	Untreated
3	10	24	5	5.0, hybrid[a] and 0.25	Flat	2	Untreated
4	10	24	5	5.0	Flat and Plunger	2	Untreated
5	10	50	5	0.25	Flat	1 and 2	Untreated
6	10	50	5	0.25	Plunger	1	Treated and untreated

[a]Hybrid fibre size is 5% for each 5-mm and 0.25-mm fibre sizes.

2.5. *Testing and Characterisation*
All the samples were tested for density, water absorption and tensile test. The fractured surface of tensile tested samples was further analysed using a polarised optical microscope.

2.5.1. *Density Test.*
The density test was referred to ASTM C29/C29M-09 [4] that used deionised water as an immersion medium using Shimadzu UX 450S densitometer. Minimum of five (5) samples with dimension 2 cm length by 2 cm width with 0.3 cm thickness were used. The samples density was obtained by measuring the samples dry weight and its volume by placing onto the upper and lower pan respectively.

2.5.2. *Voids Content Test.*
The voids content was obtained from the difference between the theoretical density and experimental density as done by Fiore, et al. [5]. The theoretical density can be calculated from equation (1).

$$\rho_t = \left(\left(\rho_m \times M_f\right) + \left(\rho_f \times F_f\right)\right) \tag{1}$$

Where, ρ_m (unit: g/cm^3) is the epoxy density, and M_f (unit: %) is the fraction of epoxy matrix. Whereas, ρ_f (unit: g/cm^3) is the density of the Kenaf fibres, and F_f (unit: %) is the fraction of Kenaf fibres in a solid composite system (no voids formation). The percentage of voids content can be seen in equation (2).

$$U_v = \frac{\rho_t - \rho_r}{\rho_t} \times 100\% \tag{2}$$

Where, U_v (unit: cm^3) is the volume of void content, ρ_t (unit: cm^3) is the samples theoretical densities, and ρ_r (unit: cm^3) is the samples experimental density [6].

IOP Conf. Series: Materials Science and Engineering **839** (2020) 012012 doi:10.1088/1757-899X/839/1/012012

2.5.3. Water Absorption Test.
The test was referred to ASTM D570-98 [7] of 24 hours of water uptake. Minimum of five (5) samples with dimension 2 cm length by 2 cm width with 0.3 cm thickness were cut and immersed in the testing cup containing deionised water. The weighing process was conducted until a plateau reading obtained. The percentage of water uptake was calculated as in equation (3).

$$\text{Water absorption (\%)} = \frac{(W_1 - W_0)}{W_0} \times 100\% \quad (3)$$

Where, W_1 is the sample mass (unit: gram) after 24 hours immersion, W_0 is sample mass (unit: gram) before the test immersion.

2.5.4. Tensile Test.
The tensile test was conducted using the Shimadzu AG-X4205 universal testing machine based on ASTM D3039/3039M-08 [8]. Five (5) samples with a dimension of 15 cm in length x 2 cm in width x 0.3 cm thickness were tested at a crosshead speed of 5 mm/min with 8 cm gauge length. The tensile stress, strain and modulus were calculated as in equation (4) to (6).

$$\text{Tensile stress, } \sigma_T\ (N/mm^2) = \frac{F}{A} \quad (4)$$

Where, F (unit: N) is the force that applied to the sample in the axial direction and A (unit: mm^2) is the cross-sectional area of the sample.

$$\text{Tensile strain, } \varepsilon_T\ (\%) = \frac{L_1 - L_\circ}{L_\circ} \times 100 \quad (5)$$

Where, L_1 (unit: mm) is the samples length after deformation and $L_\circ$ (unit: mm) is the initial length of the sample.

$$\text{Tensile modulus, } E_T\ (N/mm^2) = \frac{\sigma}{\varepsilon} \quad (6)$$

Where, E_T is the tensile modulus, σ is the tensile stress and ε is the tensile strain.

2.5.5. Characterisation of Composites Morphology.
The fractured surface of the tensile tested sample was analysed using darkfield in reflected light. The voids formation inside the composite system were analysed using brightfield in transmitted light via polarised optical microscope Olympus BX 51 as referred to ASTM F728-81(1997)e1 [9].

3. Result and Discussions

3.1. Analysis 1 (Curing temperature of pristine epoxy)
Figure 2 shows the tensile properties of pristine epoxies that fabricated at various curing temperatures. For the tensile stress as in figure 3 (a) and (b), the epoxy that cured at 24 °C temperature recorded the lowest tensile stress and modulus properties. It might be due to the presence of voids inside its system. The high viscous epoxy might reduce the mobility of bubbles to escape from its system thus creating more voids once harden and reduce its tensile stress. In addition, due to the low-stress transfer as the result of high voids formation, the 24 °C sample has low resistance to break and lowering its tensile modulus.

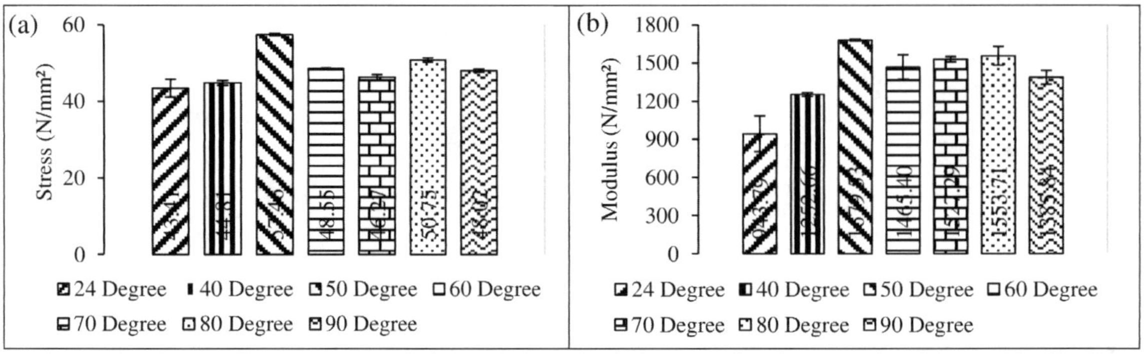

Figure 3. The tensile (a) stress and (b) modulus of pristine epoxies that cured at 24, 40, 50, 60, 70, 80 and 90 °C moulding temperatures.

By comparing the tensile stress and modulus pattern, an increasing trend was observed and reduced slightly at 90 °C sample. It might be due to the effect of high curing temperature on the voids mobility to efficiently aerate out from the 90 °C sample matrix. Since the high curing temperature accelerates the epoxy molecule to create faster 3-Dimensional links thus reduce the fabrication time that needed for voids escape out from the matrix system. Thus, this drops slightly the stress and modulus for 90 °C sample. But, for the 50 °C sample, it shows the highest tensile stress and modulus compare to the other samples. It might be because it is an ideal temperature for voids to efficiently escape out from its system thus contribute to an increment of its tensile stress and modulus properties.

3.2. Analysis 2 (Matrices mixing time)

Figure 4 shows the comparison of optical micrographs on the surface and inside the 10 wt% of the fibre reinforced composites. For 24 °C samples, as the mixing time increasing, the amount of voids also increased. The voids formation might be affected by the workable time of the epoxy matrix. As the mixing time increased, more air being whipped into the matrix thus, reducing the chance of voids to effectively escape-out from the matrix system. In addition, the size of the voids is also getting smaller as the mixing time increased since the increasing number of voids restrict their movement out from the matrix.

IOP Conf. Series: Materials Science and Engineering **839** (2020) 012012 doi:10.1088/1757-899X/839/1/012012

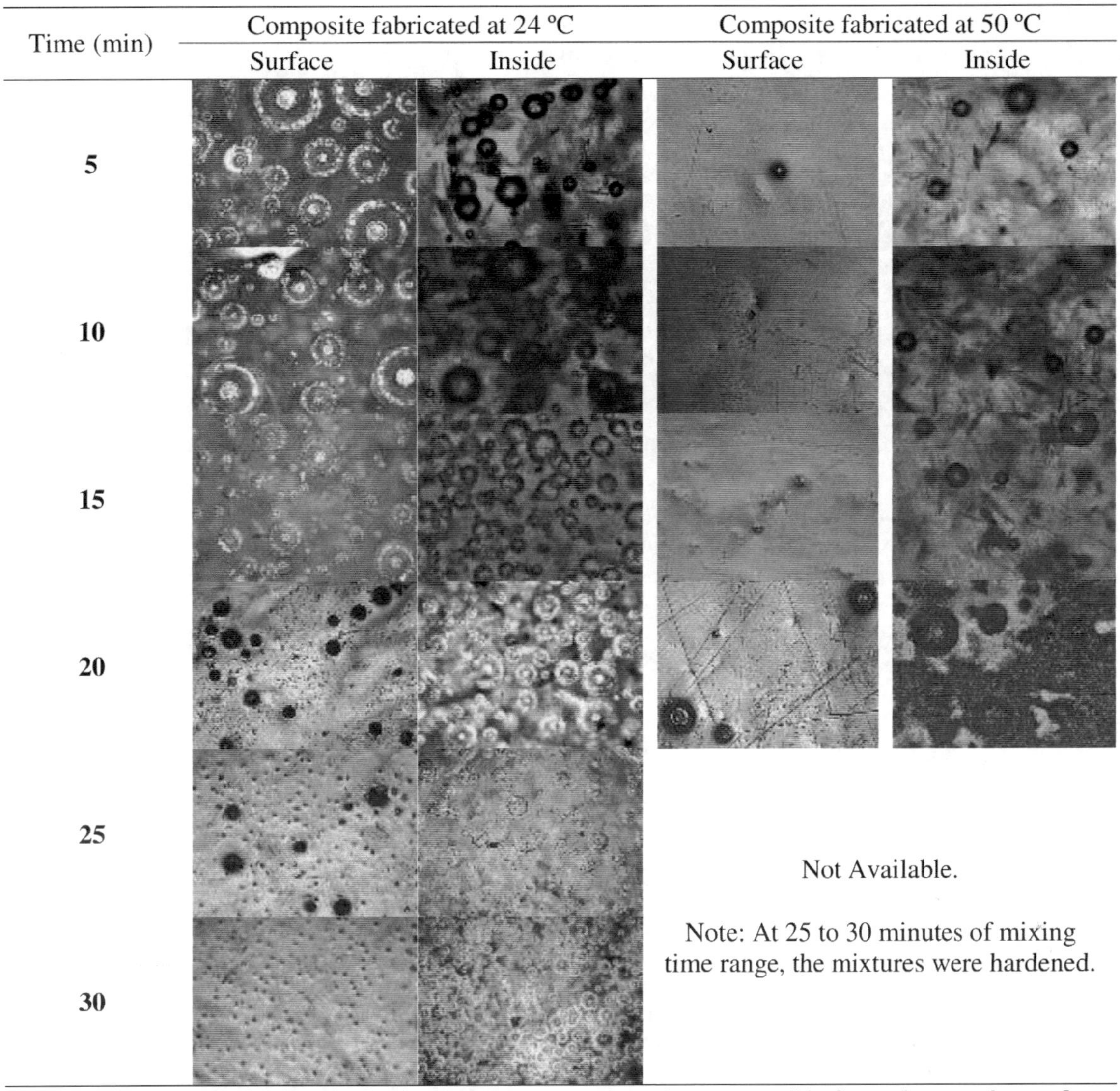

Figure 4. The polarised optical micrographs comparison between voids formation on the surface and inside the 10 wt% filled untreated fibres in the epoxy matrix composite. Fabricated at 24 °C and 50 °C curing temperatures. Captured via bright and darkfield in reflected and transmitted light at 5x magnification.

For samples the 50 °C samples, the same pattern was observed since more voids formed on and inside the matrices system as its mixing time increases. However, it noticed that the formation of the voids is less than the 24 °C samples. It might be due to the effect of heat that applied to the mixture since it results in the reduction of viscosity of the mixtures thus promote to the voids release from its system. For the 50 °C workable time, by heat application, it shortens to 20 minutes compare to the 24 °C sample.

The heat might accelerate the reaction time for the formation of 3-D links in the composite system and reducing its 'pot life'. Therefore, by comparing between mixtures, 50 °C sample at 5 minutes of mixing time appeared to has a better morphological property as it observed having low in voids formation as compared to the other samples. It might be as the effect of the optimum mixing time that slowly permeates out the from the composite system compare to the 10, 15 and 20 mixing time. Figure 5 shows the fractured surface of 24 and 50 °C samples.

Untreated fibre fabricated using 24 °C temperature		Untreated fibre fabricated using 50 °C temperature	
(a)	(b)	(c)	(d)

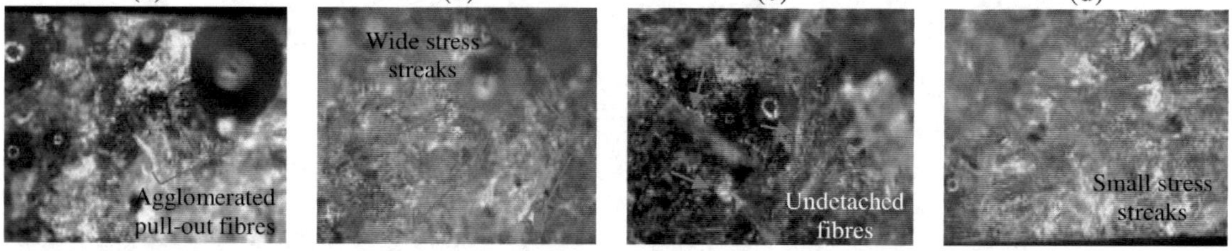

Figure 5. The fractured surfaces of the tensile tested samples that fabricated at 24 °C and 50 °C.

By comparing the condition of the fibres, the 24 °C sample (Figure 5 (a)) appeared to have fibre pull-out which indicated that low fibre-matrix interaction in the sample. The fibre also agglomerated since it showed that 24 °C sample unable to disperse well in the matrix system. Vice versa for 50 °C sample (Figure 5 (c)) that shows few untreated fibres that undetached and well dispersed throughout the region. The good distribution of stress also can be seen as the 50 °C sample have smaller stress streaks compared to the 24 °C sample as shown in Figure 5 (b) and (d).

3.3. Analysis 3 (Fibres size)
Table 4 shows the physical and mechanical properties of composites using various sizes of untreated fibre. As expected, the 10F5.0 sample recorded the lowest properties compared to the other samples. It might be due to the effect of fibres length since the long fibres might tend to agglomerate thus it inclined to create intratow voids on the unwetted fibre surfaces. Thus, it reduces its physical and mechanical properties.

For 10F0.25 sample, it shows better properties compare to the rest of the samples. It might be due to the effect of the shorter length of the fibre since it tends to dispersed well in the composite system [10]. Thus, it reduces the formation of intratow voids that caused by the agglomerated fibres. The short fibres aligned themselves in a pack formation thus better stress distribution throughout the region of the samples. As for 5FHyb sample, a slight improvement also observed in physical and mechanical properties by the combination of both 0.25mm and 5.0 mm fibres as compared to 10F5.0 sample.

It might be caused by an improvement in stress distribution as the 0.25mm fibre fill the empty gap in the composite sample to disperse the stress to the adjacent 5.0mm long fibre. Thus, this might be the explanation of the improvement in its stiffness of the 5FHyb sample.

Table 4. The physical and mechanical properties of 10 wt% of untreated fibre composites fabricated using 0.25mm, hybrid and 5.0mm fibre sizes coded 10F0.25, 5FHyb and 10F5.0 respectively.

Samples	Density (g/cm^3)		Voids Content (%)		Water Absorption (%)		Tensile Properties (N/mm^2)			
	Average	Std. Error	Average	Std. Error	Average	Std. Error	Stress	Std. Error	Modulus	Std. Error
10F0.25	1.09	0.004	6.04	0.02	3.62	0.78	30.91	1.85	1313.57	199.32
5FHyb	1.04	0.001	10.63	0.01	5.04	0.24	27.76	0.55	1544.80	33.97
10F5.0	0.95	0.013	18.60	0.04	13.19	0.47	20.33	0.20	1188.38	30.04

Note: Density of pristine epoxy at 24 °C moulding temperature is 1.16 ± 0.002.

3.4. Analysis 4 (Types of mould)

Table 5 shows the physical and mechanical properties of 5 mm fibres that fabricated using both flat-top and plunger-type mould. It was observed that the 5 mm fibre that known have lower properties improved by the usage in the plunger-type mould. It might be influenced by the constant load and temperature applied by the thick plunger mould. It suppressed the formation of the voids that formed inside the composite system by efficiently aerate-out the voids from the mould cavity. Thus, it improves the composites physical and mechanical properties.

Table 5. The physical and mechanical properties of 10 wt% of untreated 5.0 mm fibre composites fabricated using flat and plunger-type moulds coded 10FFlat and 10FPlun respectively.

Samples	Density (g/cm^3)		Voids Content (%)		Water Absorption (%)		Tensile Properties (N/mm^2)			
	Average	Std. Error	Average	Std. Error	Average	Std. Error	Stress	Std. Error	Modulus	Std. Error
10FFlat	0.95	0.002	18.60	0.04	13.19	0.47	20.33	0.20	1188.38	30.04
10FPlun	0.98	0.002	15.91	0.03	6.04	0.13	26.30	0.87	1232.64	141.5

Note: Density of pristine epoxy at 24 °C moulding temperature is 1.16 ± 0.002.

3.5. Analysis 5 (Mixing methods)

By referring to Table 6, it is the physical and mechanical of 0.25 mm fibre fabricated using mixing method 1 and 2. It was observed that, by using mixing method 1, the physical and mechanical properties of the composite (10F50(1)) is better than the composite using method 2 (10F50(2)). The thinning of the epoxy due to the application of heat contribute to the better surface contact onto the surface of the fibre. Thus, leads to reductions in voids and water uptake and improvement in tensile properties.

Table 6. The physical and mechanical properties of 10 wt% of untreated 0.25mm fibre composites fabricated at 50 °C using mixing method 1 and 2 coded 10F50(1) and 10F50(2) respectively.

Samples	Density (g/cm^3)		Voids Content (%)		Water Absorption (%)		Tensile Properties (N/mm^2)			
	Average	Std. Error	Average	Std. Error	Average	Std. Error	Stress	Std. Error	Modulus	Std. Error
10F50(1)	1.17	0.001	1.20	0.001	1.30	0.02	56.58	0.23	1667.42	28.73
10F50(2)	1.15	0.002	3.67	0.005	2.91	0.12	46.04	0.78	883.73	23.31

Note: Density of pristine epoxy at 50 °C moulding temperature is 1.17 ± 0.001.

3.6. Analysis 6 (Fibres treatment)

Table 7 shows the physical and mechanical properties of untreated, treated NaOH/Sil and coated 1:4, 1:5 and 1:6 samples. It is observed that the physical and mechanical properties of treated and coated samples is higher than the untreated sample. It might be due to the presence of pectin and lignin that contribute to the incompatibility between fibre and matrix [11] for 10FUn sample. The result followed by the low in properties by the 10F1:4 and 10F1:5 samples.

It might be influenced by the coating behaviour since the coating only coats at the outer part for 10F1:4 fibres whereas at half part of the 10F1:5 fibres [12]. Since, the coating was deteriorated due to the effect of the pulverisation process that resulting in a reduction of its fibres size [13]. Therefore, the coated fibre in both samples exposed the surface of the uncoated natural fibre thus resulting in reduction of its properties.

Vice versa for 10F1:6 sample since the high penetration of coating throughout the fibres structure contribute to its stiffness [12]. Thus, it reduces the samples deformation as better surface contact as it can be seen in the improvement in its density as well as low in voids content that leads in low in water uptake. The 10FTr also showing good physical and mechanical properties as it might be the effect of the chemical treatment coalition as agreed by Ghaztar [10].

The NaOH treatment is able to increase the surface interlocking and making the fibre stiff as it converts the cellulose I to cellulose II to become high in crystalline region [14]. With the combination of APS treatment, the fibre surface able to create good surface contact since the said treatment able to create the chemical link on the fibres surface to attach the different polarity between matrix and fibre. Thus, it increases the compaction and leads to reduce in voids formation, water uptake and improvement in tensile properties.

Table 7. The physical and mechanical properties 10 wt% filled composites fabricated at optimised parameters using untreated, treated and coated 1:4, 1:5 and 1:6 fibres coded 10FUnt, 10FTr, 10F1:4, 10F1:5 and 10F1:6 respectively.

Samples	Density (g/cm^3)		Voids Content (%)		Water Absorption (%)		Tensile Properties (N/mm^2)			
	Average	Std. Error	Average	Std. Error	Average	Std. Error	Stress	Std. Error	Modulus	Std. Error
10FUn	1.16	0.002	2.82	0.02	1.50	0.03	34.47	0.37	1639.91	53.63
10FTr	1.18	0.001	1.19	0.01	1.37	0.04	39.29	0.12	1663.02	39.00
10F1:4	1.17	0.001	1.43	0.01	1.47	0.04	35.88	0.29	1795.62	41.64
10F1:5	1.18	0.002	1.17	0.02	1.35	0.02	36.06	0.17	1942.08	18.76
10F1:6	1.18	0.002	1.08	0.04	1.34	0.02	42.57	0.65	2056.09	11.73

Note: Density of pristine epoxy at 50 °C moulding temperature is 1.17 ± 0.001.

4. Conclusion

All the optimisation parameters of the thermoset plastic using the press-moulding fabrication method were studied. It is observed that, better understanding of the fabrication parameters able to improve the properties of the short fibre-based biocomposites. With the analysis on fabrication temperatures, mixing time, fibre size, types of mould, mixing methods and treatments that afflicted to the fibre is crucial to get the optimised properties of a simple press thermoset sample. The properties improvement can be seen as the composites were fabricated at 50 °C, in 5 minutes, using plunger-mould type, 0.25-mm shorter fibre size that use mixing method 1 and coated fibres. The improvements are in better fibre compaction that suppress the formation of the voids which leads to low water uptake that parallels with the improvement in its stress dispersibility and retaining ability throughout the region of the samples. In summary, better understanding on the materials characteristic and fabrication process are crucial in order to obtain the optimum performance of the bio-composites.

5. References

[1] Abdollah M F, Shuhimi F F, Ismail N, Amiruddin H and Umehara N 2015 *J. Materials & Design* **67** 577-82
[2] Gurunathan T, Mohanty S and Nayak S K 2015 *Comp. Pt A: App. Science and Manufacturing.* **77** 1-25
[3] Oriental-Option S B (2010, 29 December) All-purpose epoxy http://orientaloption.com/epoxy_all_purpose.html
[4] Standard A S (C29/C29M-09) 2009 (Pennsylvania: ASTM International)
[5] Fiore V, Bella D G and Valenza A 2015 *Comp. Pt. B: Engineering.* **68** 14-21
[6] Mallick P K 2007 *Fiber-reinforced Composites: Materials, Manufacturing, and Design.* (Florida: CRC press)
[7] Standard A S (D570-98) 2010 (Pennsylvania: ASTM International)
[8] Standard A S (D3039/ 3039M-08) 2008 (Pennsylvania: ASTM International)
[9] Standard A S (F728-81(1997) e1) 2008 (Pennsylvania: ASTM International)

[10] Ghaztar M M M, Ibrahim N N I N and Romli A Z 2017 *AIP Conf. Proceeding* **1901** 30001 (New York: AIP Publishing)
[11] Ku H, Wang H, Pattarachaiyakoop N and Trada M 2011 *Comp. Pt. B: Engineering* **42** 856-73
[12] Ghaztar M M M, Ibrahim, N N I N, Zakaria S and Romli A Z 2019 *Scientific Research J.* **16** 43-58.
[13] Ghaztar M M M and Romli A Z 2018 *AIP Conf. Proceeding* **1985** 30014 (New York: AIP Publishing).
[14] Abdul K H P S, Bhat A H and Ireana Y A F 2012 *J. Carb. Polymers.* **87** 963-79

Acknowledgement
The author is gratified for the financial support by the Institute of Research Management and Innovation (IRMI), (600-IRMI 5/3/GIP (002/2018)). The author also would like to thank Polymer Composites Research Laboratory, Centre of Polymer Composites Research and Technology, Institute of Science (IOS), UiTM for the calibrated facilities.

Study of The Effect of The Electric Current Strength of The Weld on The Tensile Strength of The Welded Gas Cylinder Material Joint Capacity of 3 Kg

Usman Sudjadi, Basuki Wibowo and Erwan Hermawan
Center For Nuclear Fuel Technology – BATAN
PUSPIPTEK Area– Tangerang 15314, Indonesia

Abstract. The effect of welding electric current strength on the tensile strength of the 3 kg capacity gas cylinder material welding joint has been studied. The welding carried out in this study is a damping arc welding with variations in electric current of 40A, 60A, 80 A, 100 A, and 120 A. The results show that: The maximum tensile stress is found in the welded specimen with an electric current of 40 A with a tensile strength of 13 kg/mm^2, 60 A with a tensile strength of 12 kg/mm^2, 80 A with a tensile strength of 28 kg/mm^2, 100 A with a tensile strength of 46 kg/mm^2. The maximum tensile stress is the test object which is welded with 120 A electric current with a tensile strength of 39 kg/mm^2. With the above results, it can be seen that the highest tensile strength of maximum is the sample that is welded with 100 A electric current. The highest maximum tensile strength reaches 46 kg/mm^2. The results of observations of micro structures in welded and HAZ matrix areas are Ferrite and Perlite.

Keywords: Welding, Electric Current, Material, 3 Kg Gas Cylinder, Tensile strength, microstructure.

1. Introduction

As it is known that the Government of Indonesia in 2014 has decided that kerosene fuel is replaced with gas because the availability of kerosene in Indonesia has been greatly reduced. As we know Indonesia has large natural gas reserve with total production reached until 7,764 MMSCFD. Thus, housewives in Indonesia, since then, cook no longer use kerosene stoves. But housewives cook a lot using a gas stove with a 3 kg LPG gas cylinder [1-6]. But the reality is that many people are complaining now that the LPG gas cylinders with a capacity of 3 kg are less secure and this can pose a high risk, such as the explosion of LPG gas cylinders and some supporting components whose quality is doubtful.

One way to overcome the dangers of 3 kg capacity LPG gas cylinders is to improve the quality of cylinders and find out how the industrial management process of making 3 kg capacity LPG gas cylinders, as well as research on welding testing methods that are in accordance with established procedures. One of the most important processes in the metal, machining and manufacturing industries is the metal joining process called welding process. The welding procedure looks very simply, but actually there are many problems that occur in the field when the welding process is carried out [1-7]. The 3 kg capacity gas cylinder is a pressure vessel which is a storage place for LPG (Liquid Petroleum Gas) with carbon steel plate material while SG 295 with 2.5 mm plate thickness. In addition, the 3 kg capacity gas cylinder has three general parts consisting of the handle of the tube, the tube body, and the tube leg with a tube height of ± 83 mm and the outer diameter of the tube body is

IOP Conf. Series: Materials Science and Engineering **839** (2020) 012013 doi:10.1088/1757-899X/839/1/012013

260 mm. In this scientific work will be reported on, the study of the effect of the strength of the electric current welding on the tensile strength of the material connection of 3 kg LPG gas cylinders.

2. Experiments

The samples used in this study are material from 3 kg LPG gas cylinders (Figure 1). The first step is to open the bolt that binds to the 3 kg capacity LPG gas cylinder with the aim of removing the gas content until it runs out, to ensure the gas content contained in the 3 kg capacity LPG gas cylinder runs out, then the next process of the tube is immersed in a water bath with a time range ± 30 minutes, so that the smell of gas caused completely disappear, then the tube dried in the hot sun, and then the tube is cut with a grinding wheel. The material from the tube is cut into two parts, one part, the tube is cut into 10 samples that have been cut into rectangles according to the dimensions of length 100 mm, width 60 mm and 2.5 mm plate thickness (Figure 2). Tube plates that have been cut and then every two plates are welded with the SMAW (Shielded Metal Arc Welding) method with the variation of current strength used is 40 A, 60 A, 80 A, 100 A, and 120 A with a constant voltage of 24V. Samples after welded into 5 samples and ready to be in the stress and strain test, And the micro structure is observed.

Below, this is the original form of a 3 kg capacity gas cylinder along with its construction part as shown in Figure 1.

Figure 1. Construction of 3 kg capacity LPG gas cylinders

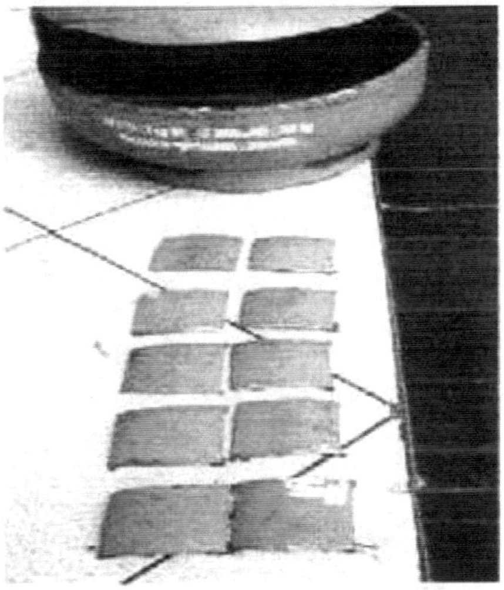

Figure 2. LPG gas cylinders 3 kg capacity are cut into 10 samples

3. Results and discussion

3.1 Tensile Test Results

From the results of tensile testing data obtained the plate tensile strength and fracture time and variations in electric current of 40A, 60A, 80A, 100A, 120A, and 140A at the welding connection, Heat Input (19.2, 29, 38.4.50, 58 and 67 kJ/cm) to get the maximum tensile stress and changes in cross-sectional area. To find out the strength of each test object, the calculation is performed in accordance with the existing data and formula. The test results are then implemented to obtain the mechanical properties of tensile testing of each test object, such as tensile strength, and changes in cross-sectional area

Table 1. Initial Tensile Strength.

Test Specimens	Sample Code (A)	Tensile Strength (kg / mm²) [Mpa]
1	40	13 [127]
2	60	12 [118]
3	80	28 [274]
4	100	46 [451]
5	120	39 [382]

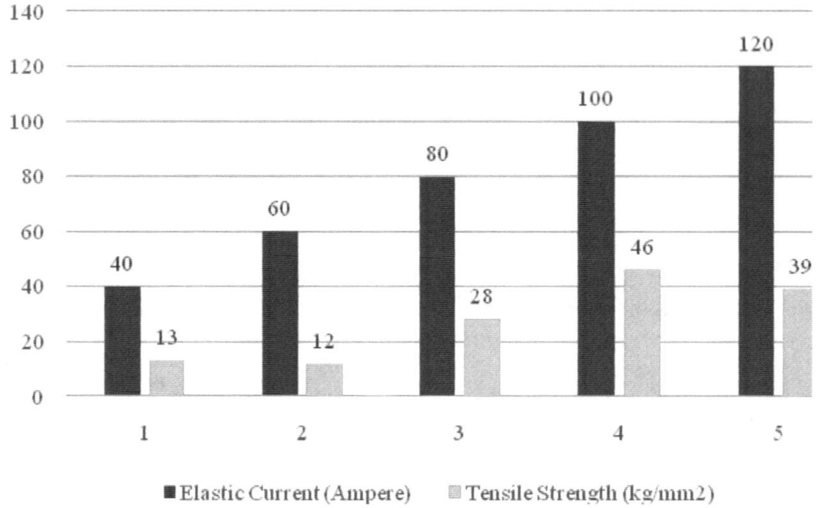

Figure 3. Graphic comparison of electric current to tensile strength (Dark blue colour is electric current (Ampere), and light bluecolour is Tensile Strength (kg/mm^2) or [MPa].

Analysis of the magnitude of the current strength and structure with a variation of the current strength of 40A, 60A. 80A, 100A, 120A, and tensile strength describe the comparison value of the tensile test of 13 kg/mm^2, 12 kg/mm^2, 28Kg/mm^2, 46 kg/mm^2, 39 kg/mm^2 from the result of tensile strength of 40A has decreased 60A. 12 kg/mm^2 and riding Ampere 80A increased 28 kg/mm^2 and increased 100A strength from 46 kg/mm^2 and added current to 120A decreased the tensile strength of 39 kg/mm^2 which is the best or the best tensile strength of 100A with a yield of 46 kg/mm^2.
Specific graph Structure comparison of electric current and tensile strength

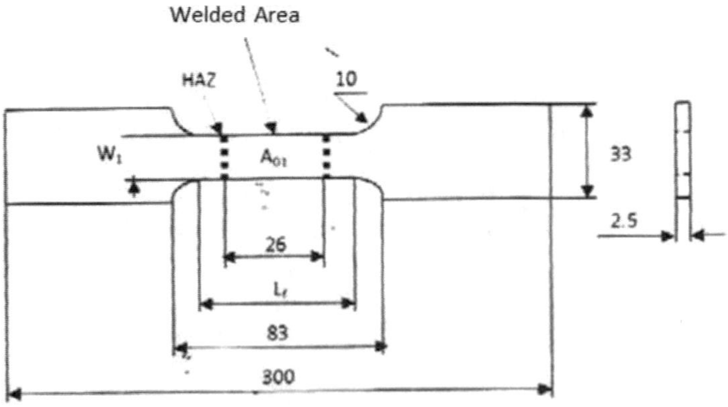

Figure 4. Tensile Test Specimens.

3.1.1 Calculation of tensile testing results.

To get the values of the mechanical properties as above, the following formulas are used:

a. The maximum tensile stress contained is held by the test object:

$$\sigma_u = \frac{F_u}{A_0} \text{ (N/mm}^2\text{)} \tag{1}$$

F_u is pull force (N) and A_o is initial area, in the middle of the tensile test sample (figure 4).

b. Strain Calculation:

$$\delta = \frac{A_1 - A_0}{A_0} \times 100\% \tag{2}$$

A_1 is area after being pulled (figure.4)

A. For Test items with sample code 40A.

a. The maximum tensile stress contained is held by the test object:

$$\sigma_u = \frac{127 \frac{N}{mm^2}}{32.25\ mm^2} = 3.93 \approx 4N \tag{3}$$

b. Strain calculation from this sample:

$$\delta = \frac{32.25 - 26}{26} \times 100\% = 24\% \tag{4}$$

B. For Test objects with sample code 60 A

a. The maximum tensile stress contained is held by the test object:

$$\sigma_u = \frac{118 \frac{N}{mm^2}}{29.25\ mm^2} = 4.03 \approx 4N \tag{5}$$

b. Strain calculation from this sample

$$\delta = \frac{29.25 - 26}{26} \times 100\% = 12.5\% \tag{6}$$

C. For Test objects with sample code 80 A

a. The maximum tensile stress contained is held by the test object:

$$\sigma_u = \frac{274 \frac{N}{mm^2}}{29.90\ mm^2} = 9.16 \approx 9N \tag{7}$$

b. Strain (δ)

$$\delta = \frac{29.90 - 26}{26} \times 100\% = 15\% \tag{8}$$

D. For the Test object with sample code 100 A.

a. The maximum tensile stress contained is held by the test object:

$$\sigma_u = \frac{451 \frac{N}{mm^2}}{29.25\ mm^2} = 15.41 \approx 15N \tag{9}$$

b. Strain (δ)

$$\delta = \frac{29.25 - 26}{26} \times 100\% = 12.5\% \tag{10}$$

IOP Conf. Series: Materials Science and Engineering **839** (2020) 012013 doi:10.1088/1757-899X/839/1/012013

E. For Test items with sample code 120 A

a. The maximum tensile stress contained is held by the test object:

$$\sigma_u = \frac{382 \dfrac{N}{mm^2}}{29.37\ mm^2} = 13.01 \approx 13N \tag{11}$$

b. Strain (δ)

$$\delta = \frac{29.37 - 26}{26} \times 100\% = 4.34\% \tag{12}$$

Table 2. Tensile test results.

No	Ampere Sample code	Dimensions (mm)	A_0 Weld-width (mm-)	W_1 Wide Cross section Welded (mm)	L_0 Initial sample Length Test (mm)	Fu force Maximum Which in Hold on Test Objects (N / mm^2)	A_1 Large Cross section End TEST objects (mm^2)	W_0 Wide Final section Test Object (mm)
1	40	13 x 2	26	12.90	300	127	32.25	33
2	60	13 x 2	26	12.72	300	118	29.25	33
3	80	13 x 2	26	13.00	300	274	29.90	33
4	100	13 x 2	26	13.00	300	451	29.25	33
5	120	13 x 2	26	13.01	300	382	29.37	33

From the test results also obtained a graph between the tensile strength and the fracture time of the recording device on the testing machine (see Figure 4-5).

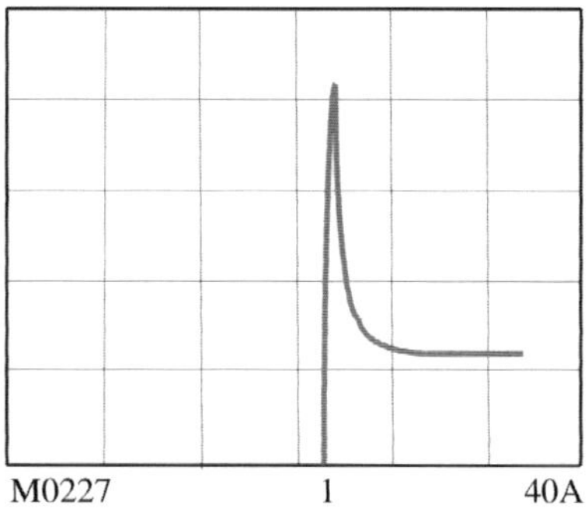

Figure 5. Graph of the results of tensile testing of 40A.

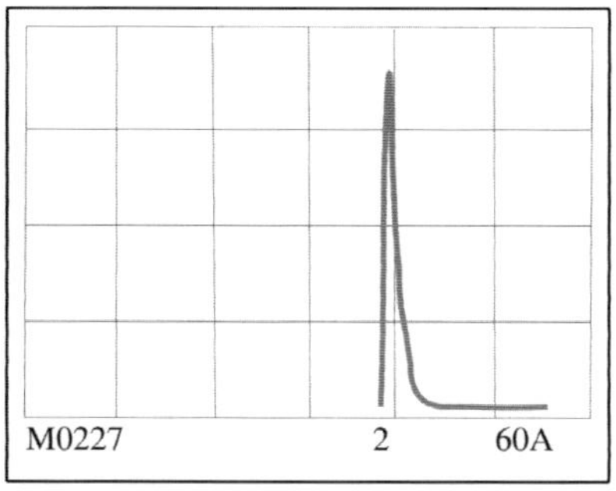

Figure 6. Graph of 60A tensile test results.

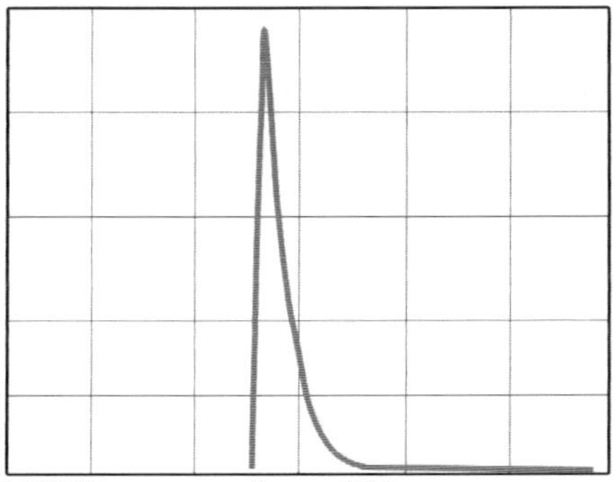

Figure 7. Graph of 80A tensile test results

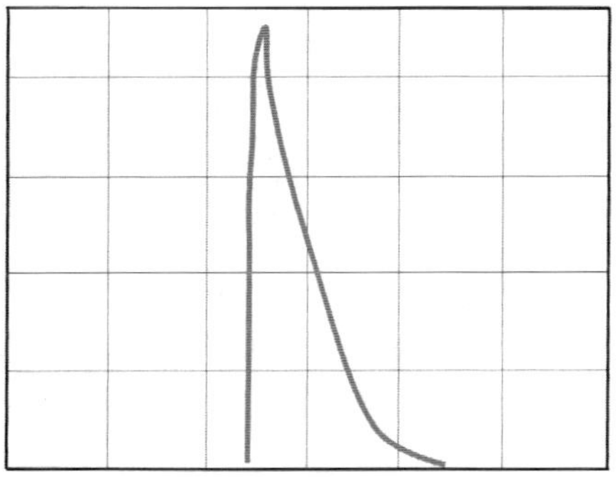

Figure 8. Graph of 100A tensile test results

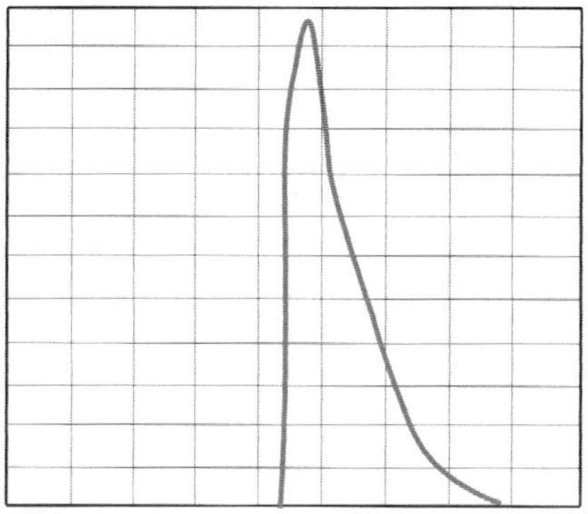

Figure 9. Graph of 120A tensile test results

In specimen 1 sample code 40A with tensile strength reaches 127 N/mm^2, specimen 2 sample code 60A with tensile strength reaches 118 N/mm^2, specimen 3 sample code 80A with tensile strength reaches 274 N/mm^2, specimen 4 sample code 100A with tensile strength reaches 451 N/mm^2, specimen 5 sample code 120A with tensile strength reaches 382 N/mm^2, with an average tensile speed of five specimens of 30 mm/min.

Table 3. Sample tensile test results

I(A)	**Stress = σ (N)**	**Strain = δ(%)**
40	3.93	24
60	4.03	12.5
80	9.16	15
100	15.41	12.5
120	13	4.34

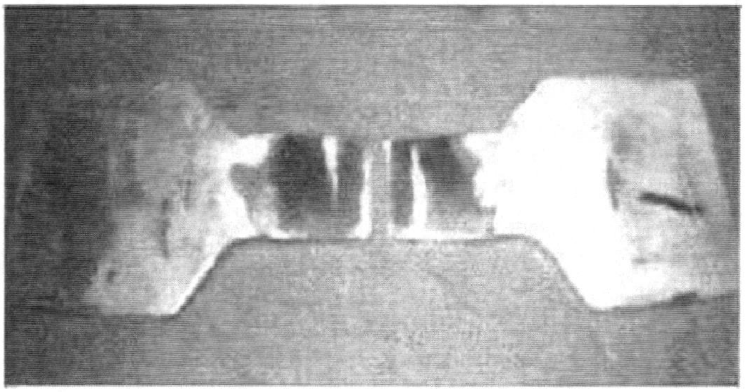

Figure 10. Sample tensile test results

In specimens 1 to 5 with variations of 40A, 60A, 80 A, 100A, and 120A, plate fractures occur at the weld joint and this concludes that at the weld joints specimen I to 5 is not perfect even though the average tensile strength value obtained for testing plates obtained prices of 27.6 kg/mm^2 [270.4 MPa] because the plates have not passed the standard test with a tensile strength of more than 41.0 kg/mm^2 recommended by SNI (Indonesia National Standard). The results are far from the SNI standard values and not broken bolts occur in the welded joint. This could be due to the influence of the parent metal, the nature of the HAZ (Heat Affected Zone) area. Weld metal properties and dynamic properties of the joint.

3.2 Micro Structure Metallographic Testing Results

The following are shown photographs of microstructure observation results of metallographic testing results with inverted metallurgical microscope olympus BX41M-LED test machine with a magnification of 500x and 1000x. This observation is to see the structure of the 3 kg LPG cylinder. Following are the results of observational photographs in the form of the spread of microstructure of metal. From observing the microstructure, it can be seen that the observed matrix is ferrite and perlite [1, 2, 9]

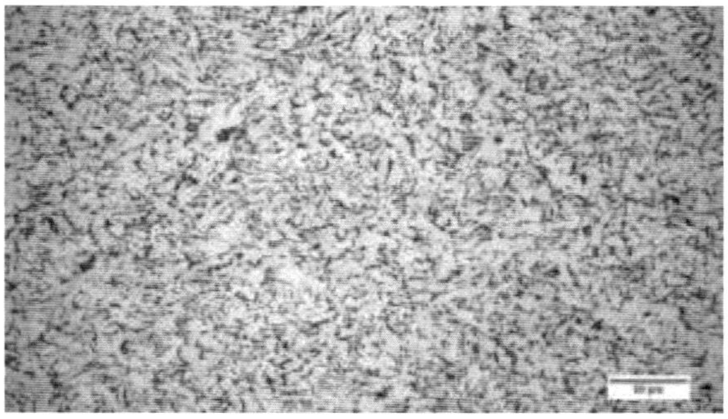

Figure 11. Photograph of microstructure observations of the material with heat input of 29 kJ/cm (60A), magnification of 500x

IOP Conf. Series: Materials Science and Engineering **839** (2020) 012013 doi:10.1088/1757-899X/839/1/012013

Figure 12. Photograph of observations of microstructure of the material by providing heat input of 38.4 kJ/cm (80A), magnification 500 times.

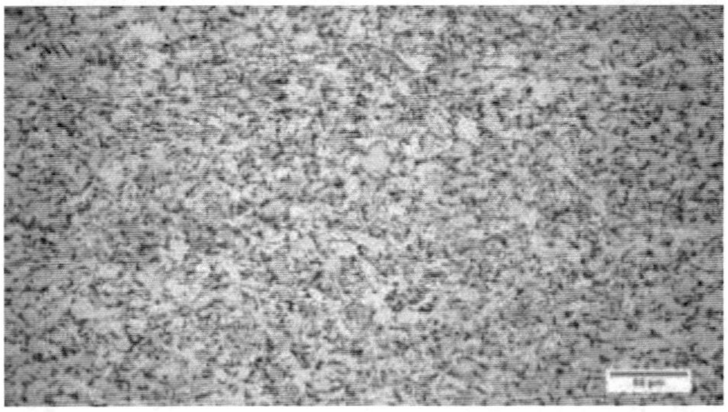

Figure 13. Photograph of microstructure observations of the material with 50 kJ/cm (100A) heat input, magnification 500 times.

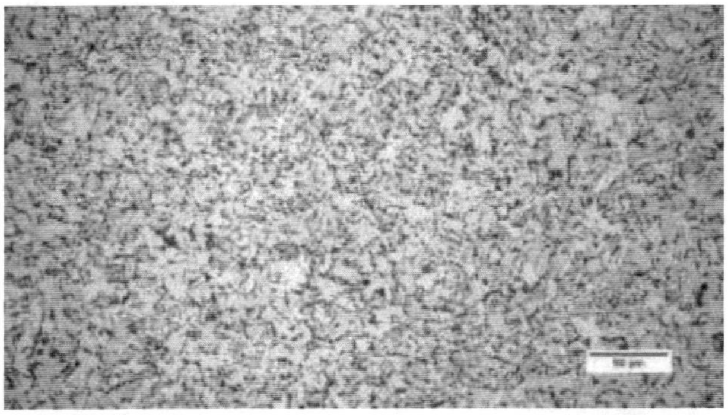

Figure 14. Photograph of microstructure observations of the material with heat input of 58 kJ/cm (120A), magnification of 500 times.

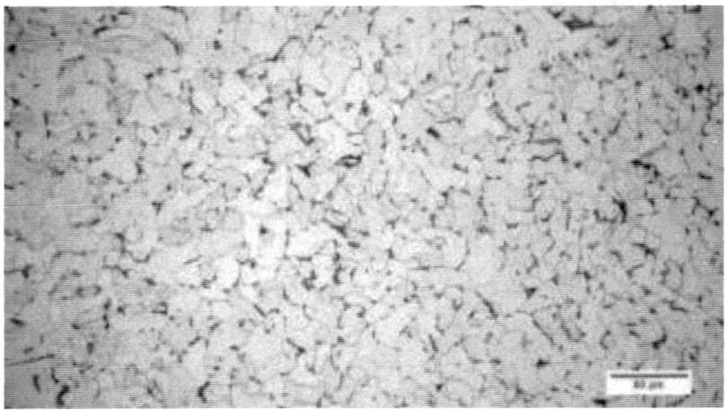

Figure 15. Photograph of microstructure observations of the material with heat input of 67 kJ/cm (40A), magnification of 500 times.

IOP Conf. Series: Materials Science and Engineering **839** (2020) 012013 doi:10.1088/1757-899X/839/1/012013

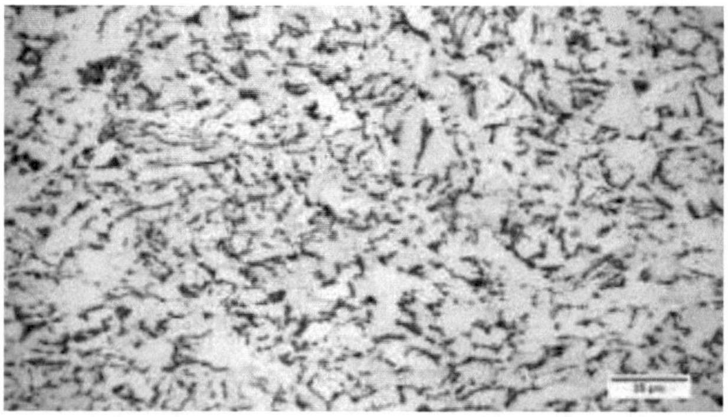

Figure 16. Photograph of observations of microstructure in the material by giving heat input of 29 kJ/cm (60A), magnification of 1000 times.

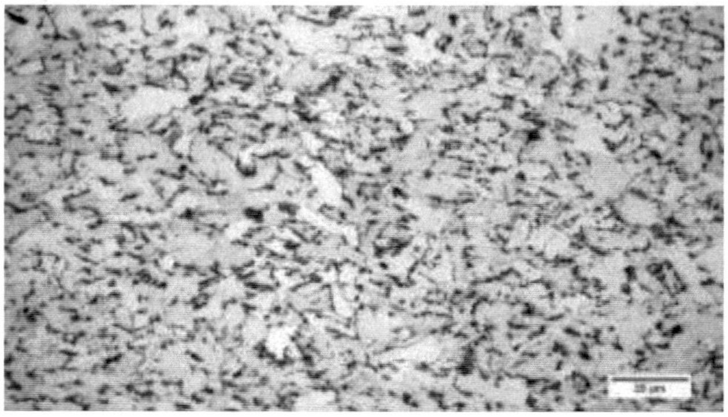

Figure 17. Photograph of microstructure observations on the material by providing heat input of 38.4 kJ/cm (80A), magnification of 1000 times.

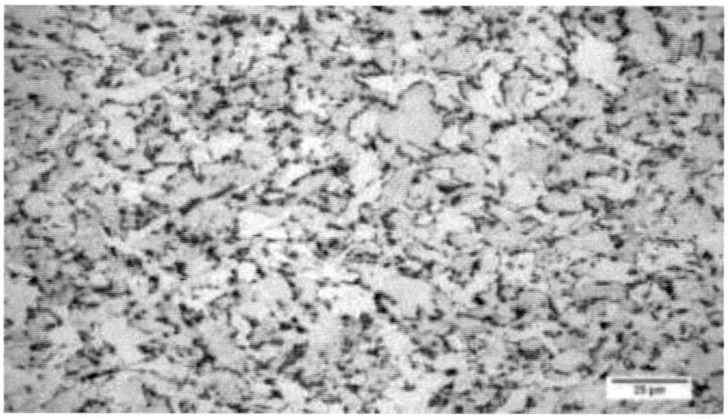

Figure 18. Photograph of microstructure observations on the material by giving heat input of 50 kJ/cm (100A), 1000x magnification, is the ferrite matrix and pearlite matrix [1,2, and 9].

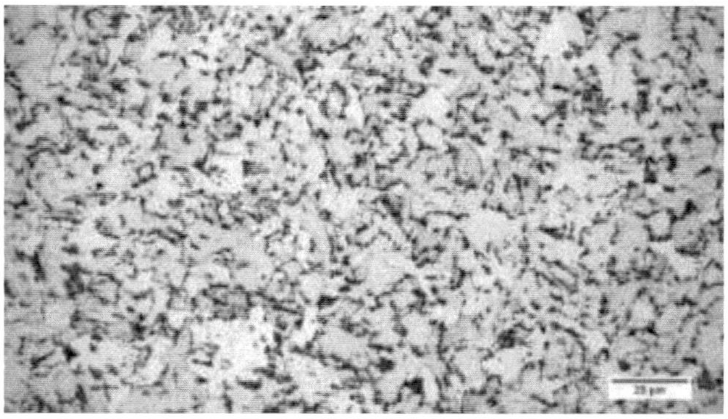

Figure 19. Photograph of observations of microstructure in the material by giving heat input of 58 kJ/cm (120A), 1000x magnification, is the ferrite matrix and pearlite matrix [1,2, and 9].

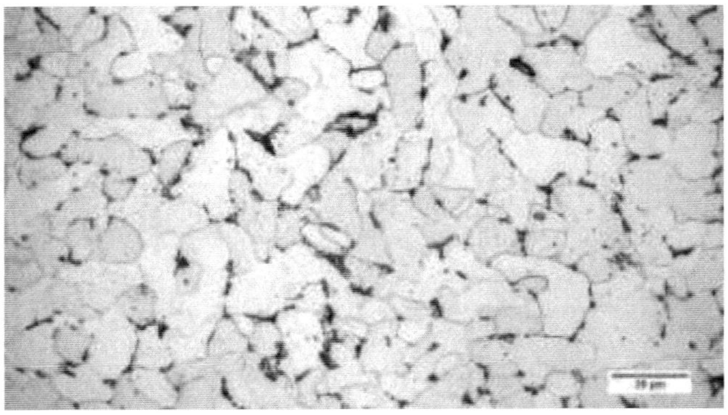

Figure 20. Photograph of observations of the microstructure of the material by giving heat input of 67 kJ/cm (40A), 1000x magnification.

Based on observations of microstructure and from reference [1,2, and 9]. So, the matrix contained in the material after welding is the ferrite matrix and pearlite matrix [1,2, and 9].

4. Conclusion

After analyzing and calculating the test result data regarding the welded joints of material from a 3 kg capacity gas cylinder, it can be concluded as follows:

Material from 3 kg LPG gas cylinders, the best / greatest tensile strength is when welded at 100 A with time (t) = 10 minutes. For samples that were welded 100 A, for time (t) = 10 minutes there was a matrix: ferrite, and pearlite [1,2, and 9]. The ferrite matrix was more even and dominant.

5. References

[1] Mulyanti and Edi S 2012 *J. Eng.* **2**
[2] Dieter G E 1990 *Mechanical Metallurgy*, Erlangga: Jakarta.
[3] Diputra R 2011 Retrieved from *http://news.okezone.com*.
[4] Anonym 2011 Retrieved from *http://www.metrotvnews.vivo.co.id*.
[5] Anonym 2006 SNI 07-3018-2006, *National Standard Information System, Indonesia*, BSN.
[6] Upstream Oil and Gas Business Activities Republic of Indonesia 2017. *Annual Report 2017.* Jakarta: Upstream Oil and Gas Business Activities Republic of Indonesia
[7] Brant D A 1985 *Metallurgy Fundamental, The Goodheradt, Willcox.*
[8] ASM Handbook 1985 *Metallography and Microstructure*, Metal Handbook Vol. **9**, American Society For Metal, USA.
[9] Barbari G 1985 SEM: *User`s Manual For MaterialScience and Engineering,* Leghigu University Bathlehem. PA 18015, Bathlehem.
[10] Cullity B D 1959. *Element of X-ray Diffraction.* Addison-Wesley. Publishing Company, INC. London.
[11] William D C, Jr. John W and Sons 2004 *Material Science and Engineering*: *An Introduction.*
[12] William N 1998 *Schum`s Outlines, Strength of Materials.*

IOP Conf. Series: Materials Science and Engineering **839** (2020) 012014 doi:10.1088/1757-899X/839/1/012014

Beneficiation opportunities for osmium: A review

Duduzile Nkomo[1] and Alain Mwamba[1]

[1]Advanced Materials Division, Mintek, 200 Malibongwe Drive, Strydom Park, Randburg 2125, South Africa.
Corresponding author: dudunk@mintek.co.za

Abstract. This review summarises the safe production of osmium and its compounds and its functionality from both the chemistry and materials perspectives. A review of the available data shows that there is a lot of potential for osmium (II) compounds to be used for traffic signalling as molecular signal gates using electrochemical generated power. Osmium (II) compounds can also be used for biomedical applications such as the development of anti-cancer drugs. Organic compounds of osmium have also been found to possess excellent electrochemical properties which are useful for redox flow battery. Different synthesis routes to produce these compounds and implementation thereof into the redox flow battery systems are described in details. This review of osmium and its compounds possess a variety of correlated properties with a potential to put an emergent demand on osmium for various applications.

1. Introduction

Osmium (Os) is one of the platinum group metals. Its reserves are mostly found in South Africa, Russia and North America. It has not been used in most applications because of its tendency to produce a dangerous compound called osmium tetroxide (OsO_4), which when inhaled it causes harm. The slow reaction of Os powder with oxygen gives off detectable amounts of OsO_4 vapour [1,2].

The OsO_4 is highly toxic with low contents of about 10^{-7} g.m^{-3} causing lung congestion, skin damage and severe eye damage. A harmful contamination of the air can be reached very quickly on evaporation of this substance at 20°C. It is a strong oxidant and it reacts with combustible and reducing materials. It is therefore necessary to come up with safe synthesis route of osmium-based compounds in order for osmium to be useful [1,2].

Osmium is commercially recovered from the platinum group metal (PGM) ores. Platinum and palladium are the most commercially used and osmium is the least used of the PGMs. South Africa, Russia and Canada have approximately 98% of known global PGM reserves. In South Africa the approximate distribution of operating costs during PGMs recovery for each stage are as follows: 72% mining, 9% smelting and 9% refining. It has been reported that osmium has been left piling in the tailing dams and waste streams. This project will focus on evaluating how much of osmium is mined as part of total PGMs and whether or not that amount can be of any use, starting with this review [3,4].

Osmium has a very high bulk modulus of 383-462 GPa, which is very close to that of diamond which is 446 GPa. Compressibility of osmium is generally low and is dependent on its pressure volume-temperature behaviour. It has a hexagonal-close packed structure. However, its mechanical and physical properties are unique. It is very hard, extremely dense and brittle which makes it difficult to process for metallurgical purposes [5]–[12]. Several properties of osmium and its compounds from both chemistry and materials perspectives have been reviewed in the current paper and the possible applications are discussed.

2. Safe synthesis of osmium-oxide compounds

Osmium tetroxide is the most useful compound of osmium. This compound is in demand for use as a catalyst for research purposes, although it is highly reactive in air and therefore dangerous. It has therefore been a difficult process to synthesis this compound and to find more suitable applications [13]. There are other osmium oxide compounds which have been explored fully by the researchers. Yamaura et al. [14] studied the high pressure and temperature synthesis of osmium oxides using belt-type high-pressure apparatus. The synthesised solid-state compounds were then characterised for consideration for possible applications.

The use of this method avoided possible man exposure to highly toxic OsO_4. Figure 1 shows the apparatus of belt-type high-pressure apparatus that were used at NIMS, Hokkaido University-Japan. The synthesis was performed by first sealing a mixture of high-purity starting material in Pt capsule in an argon filled glove box. The capsule was then statically and isotropically compressed to a pressure of 6 GPa in the belt-type high-pressure apparatus. This was followed by heating at temperature 2000°C for 1 hour [9,14].

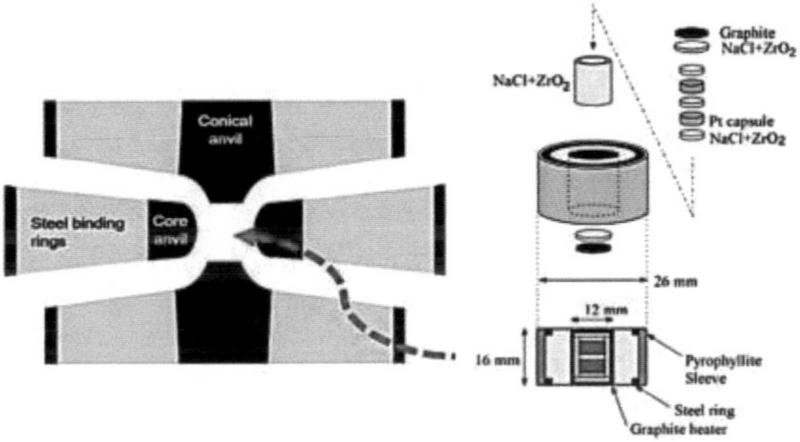

Figure 1. Assembled views of a belt-type high-pressure apparatus and sample cell [14]

The pellets obtained were then characterised using X-Ray Diffraction. His [14] results showed that there is a vast opportunity for synthesis of osmium oxides safely. Successional synthesis of Os oxides was performed, for example, $LiOsO_3$ using similar method of belt-type high pressure. However there was no success regarding the target applications which were magnetism and conductivity. These studies showed that there is a promising future for further research for Os oxide compounds and their applications [9]–[12].

3. Traffic signals

Traffic signals are developed by information programming. The chemically encoded information leads to the generation of optical output signal which in turn give rise to the development of molecular level logic gates such as AND, OR, NOR, INHIBIT, XOR, YES, NOT and XNOR. The physical integration of these molecular logic gates are related to functional integration and processing of binary data in conventional microprocessor based systems. The conventional system uses mostly silicon based technology, but miniaturisation is limited up to nanometer scale. Thus, construction of new materials that are suitable for information storage and retrieval at the molecular level have been developed to overcome the limit [15,16].

Mardanya et al. [15] utilised osmium (II) complexes based on pyrene-biimidazole ligand. This complex had remarkable photophysical properties which arise due to metal-to-ligand charge transfer transition. In order to mimic the molecular trafficking mechanism, three chemical inputs were chosen: CN^- (Input 1), H_2PO4^- (Input 2) and H^+ (Input 3). It was observed that the extent of absorption spectral changed substantially with variation of inputs. A designation was given for each output absorption signal generated at different inputs, that is, CN^- has the green signal, H_2PO4^- has the yellow signal and H^+ has the red signal [15]. This provides osmium with a promising future for the actuator applications.

4. Anti-cancer drugs discovery

Due to the chemical properties of osmium, it has been a metal of interest when it comes to research for anti-cancer compounds. Its complexes have become increasingly important in the field of tumor-inhibiting metal species in the past years. It was established that the synthesis of Os (II) complexes demonstrated that they have a comparable cancer cell cytotoxicity to cisplatin [17].

The focus in recent years has been on the development of chemotherapeutic osmium compounds, which are organometallics with arene or cyclometallated ligands. The studies showed that the arene complexes have a limitation because they are not stable in water or air. However, the problem was resolved by modulating the steric and electronic properties of the substituents on chelating ligands to control the kinetics and thermodynamics of osmium drugs in aqueous media. In addition, metal complexes of osmium in low oxidation state (+2, +3 and +4) have anti-cancer properties by inducing cell death via DNA targeting [17,18].

Recently, the first osmium complexes to induce cell death via endoplasmic reticulum stress were synthesised. Os (VI) nitride compounds bearing bidentate ligands showed that small modifications to ligands periphery induced completely different cellular responses, which lead to genomic arrests that could be chemotherapeutic [18,19].

5. Biosensors

Lithium and alkaline electrolytes batteries have long been accepted and used to power implantable and semi-implantable devices, such as insulin pumps, neurostimulators and pace makers. These batteries provide power through an internal chemical reaction. The modern technology seeks to improve these batteries by shrinking battery size and enhancing lifetime [20].

This could be achieved by developing medical device that takes advantage of fuel cells that derive their fuel and oxidant from the in-vivo environment. With such development, there will be no requirements to provide anode and cathode power generating capacity within the device, allowing miniaturisation. This therefore presents a new innovative direction in the medical device technology [20].

Osmium-based redox complexes have been found to be excellent electron transfer mediators due to their relative stability in both oxidised and reduced states and their fast electron exchange rate compared to other metal based mediators. Osmium based complexes can be widely explored for a range of applications such as redox catalysts and mediators due to ease of synthetic variation of structure. Polyvinyl imidazole (PVI) bound osmium polypyridyl series of redox polymers have been widely used in mediated enzyme electrodes for application such as biosensor and enzymatic fuel cell electrodes for power generation [20]–[22].

Bollella et al. [20] made a comparison between two plant peroxidases, cationic horseradish peroxidase (HRP) and anionic tobacco peroxidase (TOP) combined with a highly cationic osmium polymer $[Os(4,4'\text{-dimethyl-2,2'bipyradin})_2Poly(N\text{-vinylimadazole})_{10}Cl]^{+2/+}$ ($[Os(dmp)PVI])^{+/2+}$) to develop highly sensitive, stable and selective hydrogen peroxide biosensors. Furthermore, they carried out electrochemical experiments to investigate the influence of surface charge of the enzyme and the charge of the polymer on the efficiency of electron transfer between the enzyme and the wiring Os-redox polymer, see figure 2. The electron transfer rate decreased in HRP due to the repulsion between the enzyme and the polymer, both positively charged, whereas with TOP there was an enhanced electron transfer due to attraction between the anionic enzyme and the cationic polymer [20,23].

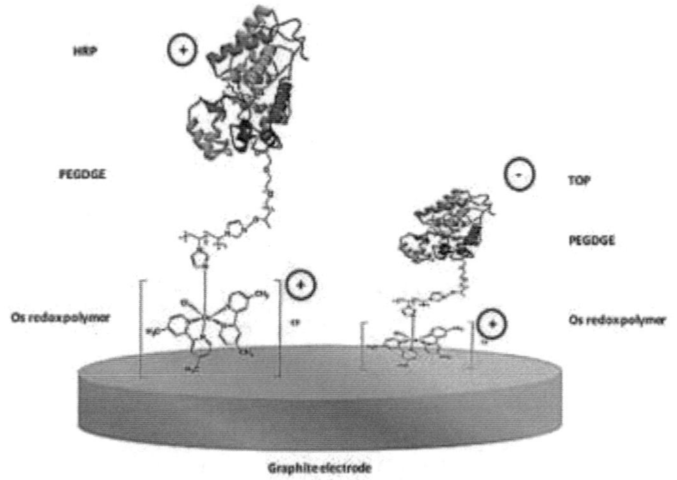

Figure 2. Schematic representation of HRP and TOP Os-polymer biosensors [20]

Synthesis of the above mentioned Os-redox polymer biosensors allowed for accurate, cost-effective determination of hydrogen peroxide by electrochemical sensing based on electron transfer while saving time. Therefore, it was concluded that the wiring of redox enzymes with Os-polymer ensured an efficient electron transfer between the prosthetic group of enzyme and the electrode surface [20]. Such properties of Os-polymer is interesting for further studies in the biomedical field for innovative development of biosensors.

6. Redox flow battery applications

Redox flow batteries (RFBs) are attractive as stationary energy storage devices because of their rapid response times, high energy capacities and high storage efficiencies. Furthermore, they are advantageous because of their ability to provide specific power and energy requirement for a specific application. The storage capacity can be increased by adding electrolyte, so the incremental cost of each additional energy storage capacity unit is reduced than in other types of batteries [24].

It has been reported that the use of purely metal-based RFBs will most likely not be the energy-storage system of choice due to usage of toxic and harmful halogens and complexing agents, high-priced metal salts, corrosive acidic electrolytes and expensive ion-exchange membranes. In the VRFB systems, vanadium contributes about 30% to the overall cost of the battery. Therefore, the higher the price of vanadium, the higher the cost of the battery [25].

These redox flow batteries limitations can be overcome by the development of environmentally sustainable, cost effective redox-active organic charge-storage materials, whose solubility and electrochemical properties can be adjusted by purposefully introducing substituents and functional groups instead of inorganic halogens. Aqueous all-organic RFB utilises low-cost membranes and sustainable, inexpensive organic-charge storage materials [25].

The development of osmium-based organic redox flow battery promises to be advantageous over other battery systems because; Organic redox flow batteries are environmentally friendly, low CAPEX is achievable because of the readily available osmium and easy synthesis of organic molecules, large capacity is achieved with electrolyte addition, and robust simple design. The organic batteries are modular and scalable because electrolytes can be replaced by new ones, more capacity with more electrolytes, and more power with more stacks. Here one has to choose whether or not power or stored energy is important, depending on the application. Low OPEX is achievable because of minimal maintenance, long life, and sustained performance [26,27].

The complexicity of osmium compounds is a promising route for improvement of redox-active materials for use in RFBs. The use of Os-complexes promises the increased voltage efficiency and energy efficiency because it meets the requirements of a bipolar redox-active material. Generally, the synthesised organic osmium compounds should be able to meet the requirements for redox-active material, that is, it should offer properties such as:

- Feature two reversible redox reactions and can be used simultaneously as both cathode and as anode material in a symmetric RFB [29]–[31]
- Provide easier synthesis route [32]–[34]
- Eliminate the use of expensive ion-exchange membranes [35]–[37]
- Must have high electrochemical reversibility [25]
- Transfer more than one electrons during the redox process [25]
- Must have high cell voltage [25]
- Must have high solubility in the applied solvent [25]

In this current review, the summary of the chemistry of organic osmium compounds has been done and the characteristics mentioned above has been reviewed and reported. The first compounds to be discussed are the 2-(phenylazo) pyridine osmium complexes. The chemistry of 2-(phenylazo) pyridine osmium complexes have been studied and reported by Das et al.[33]. The $OsBr_2$ moiety was used as the reactive fragment because the Os-Br bonds disassociate easily even in mild condition, with two π-acid ligands, that is, 2-(phenylazo) pyridine (pap, 1) and tryphenylphosphine (PPh_3) used to satisfy the remaining four coordination sites of Os in $OsBr_2$. The chemical synthesis of 2-(phenylazo) pyridine osmium complexes is explained in [33]. The synthesis allowed for the successive synthesis of [Os $(PPh_3)_2$(pap) Br_2] and [Os^{II} $(PPh_3)_2$ (pap) (L)] ClO_4 (L= bidentate ligand). The electrochemical properties of these complexes were evaluated using cyclic voltammetry and the resulting voltammograms are shown in figure 3 for [Os $(PPh_3)_2$(pap) Br_2] in dichloromethane solution and [Os^{II} $(PPh_3)_2$ (pap) (L)] ClO_4 in acetonitrile solution.

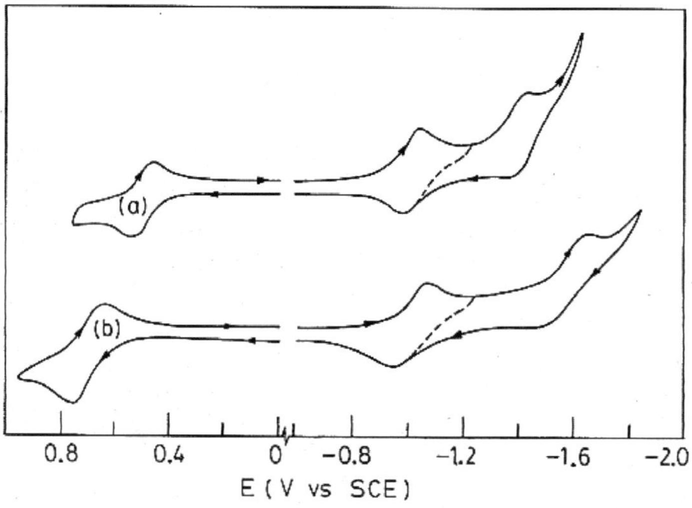

Figure 3. Cyclic voltammograms of: (a) [Os $(PPh_3)_2$(pap) Br_2] and (b) [Os^{II} $(PPh_3)_2$ (pap) (L)] ClO_4 at a scan rate of 50 mV s^{-1} [33]

The [Os $(PPh_3)_2$(pap) Br_2] showed a reversible oxidative response at 0.49 V vs. SCE and two quasi-reversible reductive responses at -1.03 and -1.40 V vs. SCE. The oxidation is assigned to osmium (II)-osmium (III) oxidation. As part of the discussions, it was stated that the 2-(phenylazo)-pyridine ligand is known to accept successively two electrons in its lowest unoccupied molecular orbital. The cyclic voltammetry of [$(PPh_3)_2$ (pap) (L)] ClO_4 also showed reversible Os (II)-Os (III) oxidation at the potential varying between 0.69 V and 1.01 V vs. SCE and two reductions below -0.9 V vs. SCE of the coordinated pap in acetonitrile solution, which has been used in RFBs as well [33]. These compounds are therefore excellent candidates for use as redox-active materials in RFBs and will be investigated further in the current project.

Another possible Os-complexes that promises to be useful in RFBs include the work that was done by Clayton et al.[38]. His work details the synthesis of arene-osmium complexes. The synthesis route of these complexes in shown in figure 4. The reactivity of Os-complexes were characterised using the NMR. However, the characterisation did not include electrochemistry. Since the synthesis of these compounds have been completed, it is worth a while to consider looking at cyclic voltammetry in order to evaluate the suitability of these complexes for use in RFBs. It was further stated that osmium has a shielding effect on the methylene protons adjacent to it. This might be the foundation for synthesis of a bipolar organic Os redox-active material to be used in RFBs [38].

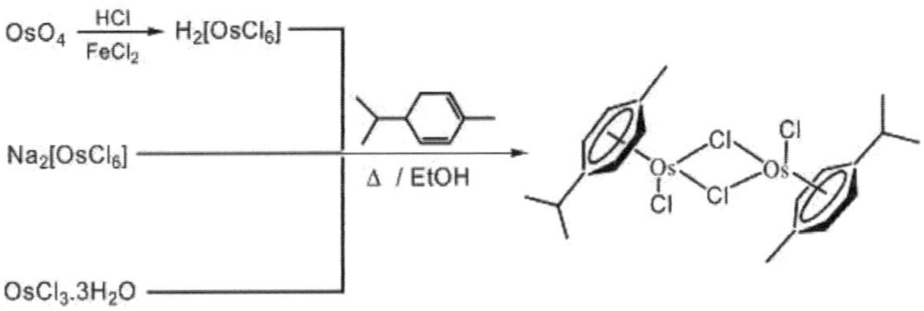

Figure 4. Schematic presentation of synthetic methods for dimeric η^6-p-cymene Os (II) complexes[38]

Another interesting Os (III) complex bearing two (carbene) borate ligands was synthesised by Chen et al.[32] building from the work done by Jia et al.[39]. The synthesis was performed by first reacting $LiPhB(MeIm)_3$ with $[Os(COD)C_{l2}]_n$, of which no observed reaction took place after 1 day at temperatures between -20 and 60°C. In the second attempt, $(Bu_4N)_2OsCl_6$ was used as a starting material leading to formation of a purple solution. The resulting solution was purified and recrystallized to give a pure crystalline sample of $[PhB(ImMe)_3]_2Os(OTf)$ (**1b**). The synthesis scheme is shown in figure 5 [32].

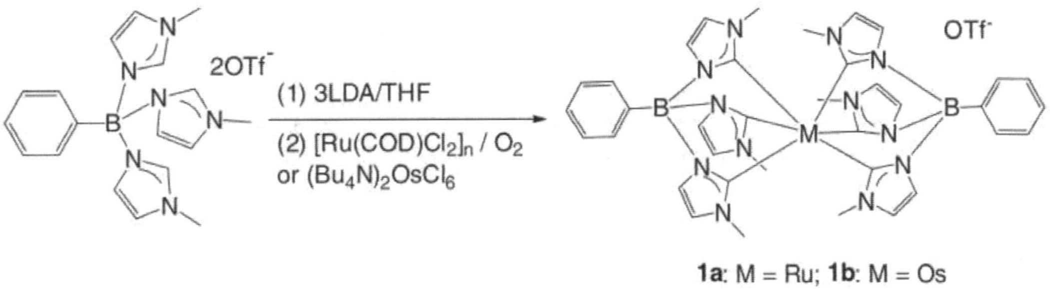

Figure 5. Synthesis route of $[PhB(ImMe)_3]_2Os(OTf)$ (1b) [32]

It was reported that during the formation of compound (**1b**), the starting Os (II) was oxidised to Os (III). The complex was stable in solid and solution state. This is one of the important factors that enhance the performance of the battery. Therefore, this compound is one of the potential compounds to be considered for use in RFBs as the stability determines the life-time of the battery. Furthermore, it was reported that no decomposition was observed after standing in air for three months [32]. Further tests were carried out using cyclic voltammetry to investigate the electrochemical properties of

$[PhB(MeIm)_3]_2Ru(OTf)$ (1a) and $[PhB(MeIm)_3]_2Os(OTf)$ (1b) in MeCN/0.1 M NBu_4PF_6, glassy carbon electrode, scan rate 100 mV/s. The results showed two reversible redox waves, as shown in figure 6.

A reversible wave with $E_{1/2}$ value at 0.35 V vs F_C^+/F_C couple was attributed to the Os^{III}/Os^{IV} couple. Another reversible wave was observed at -0.86 V and attributed to Os^{II}/Os^{III} couple. The reversible redox waves indicated the strong donating nature of the tris-(N-heterocyclic carbene) borate ligand. This means $[PhB(ImMe)_3]_2Os(OTf)$ compound is one of the potential for use in the RFB applications, which will take into advantage its electrochemical properties and easy synthesis routes [32].

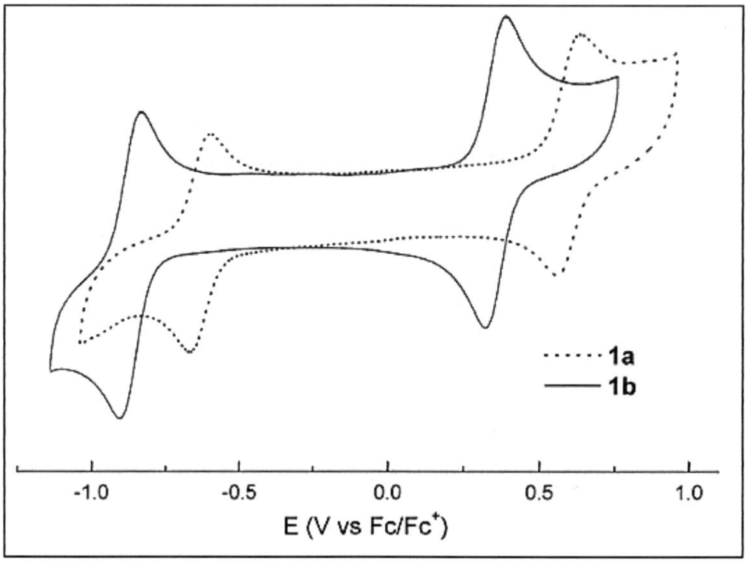

Figure 6. Cyclic voltammograms of $[PhB(MeIm)_3]_2Ru(OTf)$ (1a) and $[PhB(MeIm)_3]_2Os(OTf)$ (1b) [32]

7. Conclusions and future outlook

Over the past few years, there has been an interest in osmium for use in various applications and the initial research results suggest that it could find future application in complexes used in biosensors, biofuels and organometallic drugs for treating cancer. The establishment of easy and safe synthesis routes for Os compounds will also place a demand on osmium for use in redox flow batteries. The entry barriers to new entrants in such advanced applications, however, appears to be high, requiring access to precursor materials, strong collaborative research partnerships with universities, potential end-users, and sustainable long-term research funding.

8. References

[1] Kong S 2011 *United States Patent* **2** 12–15

[2] Li S, Ma C, Zhang Q, Liu X and Yi Z 2015 Preparation and characterization of osmium films on quartz substrate by magnetron sputtering method *Surf. Coatings Technol.* **282** 1–5

[3] Bird J, Meibom A, Frei R and Nägler T 1999 Osmium and lead isotopes of rare OsIrRu minerals: Derivation from the core-mantle boundary region? *Earth Planet. Sci. Lett.* **170** 83-92

[4] Mpinga C, Eksteen J, Aldrich C and Dyer L 2005 Direct leach approaches to Platinum Group Metal (PGM) ores and concentrates: A review *Miner. Eng.* **78** 93–113

[5] Li Y, Hao J and Xu Y 2012 Predicting hard metallic osmium-carbon compounds under high pressure *Phys. Lett. Sect. A Gen. At. Solid State Phys.* **376** 3535–39

[6] Zemzemi M and Hebbache M 2008 Ab initio investigation of osmium carbide *Int. J. Refract. Met. Hard Mater.* **26** 61–67

[7] Hebbache M, Stuparević L and Živković D 2006 A new superhard material: Osmium diboride OsB_2 *Solid State Commun.* **139** 227–31
[8] Juarez-Arellano A et al. 2009 Stability field of the high-(P, T) Re2C phase and properties of an analogous osmium carbide phase *J. Alloys Compd.* **481** 577–81
[9] Yamaura K 2016 Short review of high-pressure crystal growth and magnetic and electrical properties of solid-state osmium oxides *J. Solid State Chem.* **236** 45–54
[10] Liu K, He D, Zhou X and Chen H 2011 First-principles study of structural and thermodynamic properties of osmium *Phys. B Condens. Matter.* **406** 3065–69
[11] Zhang M, Wang M, Cui T, Ma Y, Niu Y and Zou G 2008 Electronic structure, phase stability, and hardness of the osmium borides, carbides, nitrides, and oxides: First-principles calculations *J. Phys. Chem. Solids* **69** 2096–02
[12] Yokoyama T, Walker D and Walker J 2009 Low osmium solubility in silicate at high pressures and temperatures *Earth Planet. Sci. Lett.* **279** 165–73
[13] Griffith P 1974 Osmium Tetroxide and Its Applications *Platin. Met. Rev.* **18** 94–96
[14] Feng H, Sathish C, Li J, Wang X and Yamaura K 2013 Synthesis, structure, and magnetic properties of a new double perovskite Ca_2InOsO_6 *Phys. Procedia* **45** 117–20
[15] Mardanya S, Mondal D, Karmakar S and Baitalik S 2017 Smart ruthenium and osmium complexes mimic the complicated functions of traffic signal and memory device *Sensors Actuators B Chem.* **239** 635–41
[16] Kotwal A, Lee S and Kim Y 2013 Traffic Signal Systems: A review of current technology in the United States *Sci. Technol.* **3** 33–41
[17] Babak M and Hartinger C 2014 Development of anticancer agents: Wizardry with osmium *Drug Discov. Today* **19** 1640–48
[18] Muhammad N and Guo Z 2014 Metal-based anticancer chemotherapeutic agents *Curr. Opin. Chem. Biol.* **19** 144–53
[19] Kim S, Chung J and Kim B 2011 Recycling of osmium catalyst in oxidative olefin cleavage: A chemoentrapment approach *Tetrahedron Lett***52** 1363–67
[20] Bollella P et al. 2018 Highly sensitive, stable and selective hydrogen peroxide amperometric biosensors based on peroxidases from different sources wired by Os-polymer: A comparative *Solid State Ionics* **314** 178–86
[21] Darus L et al. 2017 Redox-Polymers enable uninterrupted day/night photo-driven electricity generation in biophotovoltaic devices *J. Electrochem. Soc.***164** 3037–40
[22] Conghaile P 2013 Tailoring surfaces and supports for enzyme mediation with application to and biopower device development (*Pre print*)
[23] Bollella P et al. 2017 A Comparative Study between hydrogen peroxide amperometric biosensors based on different peroxidases wired by Os-polymer: Applications in water, milk and human urine *5th international symposium on sensor science vol 1 (*Barcelona: Spain*)* p 699
[24] Shah A et al. 2017 Recent developments in organic redox flow batteries: A critical review *J. Power Sources* **360** 243–83
[25] Jena F 2018 Organic redox-active materials for redox-flow batteries *Thesis*
[26] McHugh J 2015 Technology performance report: Vanadium redox flow battery project american recovery and reinvestment Act
[27] Chakrabarti M, Hajimolana S, Mjalli F, Saleem M and Mustafa I 2013 Redox flow battery for energy storage *Arab. J. Sci. Eng.***38** 723–39
[28] Kemwatt, 2018 Environmentally friendly: The first industrial organic redox battery systems *Off-grid power conference* (Munich) p 1-28
[29] Wedege K, Dražević E, Konya D and Bentien A 2016 Organic redox species in aqueous flow batteries: Redox potentials, chemical stability and solubility *Sci.Rep.***6** 1–13
[30] Chen H, Cong G and Lu Y 2018 Recent progress in organic redox flow batteries: Active materials, electrolytes and membranes *J. Energy Chem.***27** 1304–25

[31] Likit-Anurak K, Uthaichana K, Punyawudho K and Khunatorn Y 2017 The performance and efficiency of organic electrolyte redox flow battery prototype *Energy Procedia* **118** 54–62
[32] Chen F ,Wang W, Li Y, Chen X and Xue X 2012 Syntheses, structures and electrochemical properties of homoleptic ruthenium(III) and osmium(III) complexes bearing two tris(carbene)borate ligands *Inorg. Chem. Commun.* **21** 88–91
[33] Das A, Peng S and Bhattacharya S 2000 Chemistry of 2-(phenylazo)pyridine complexes of osmium: Synthesis, characterization and reactivities *Polyhedron* **19** 1227–32
[34] Romanov A, Muratov D and Kudinov A 2013 High-yield synthesis of osmium complexes [(η6-C 6H6)OsCl2]2, [CpOs(η 6-C6H6)]$^+$ and CpOs(η 5-cyclohexadienyl) *J. Organomet. Chem.* **724** 177–79
[35] Brushett F, Milshtein L, Su L Aqueous all-organic redox flow battery, Brushett research group, *https://www.brushettresearchgroup.org* (Cambridge MA: Massachusetts Institute of Technology)
[36] M 2017 The vanadium redox flow battery and South Africa's export opportunity: Introduction and objectives *SA Energy Storage Conf.Bushveld energy* (South Africa) *www.bushveldenergy.com*
[37] Sprenkle V et al.2012 Next Generation Redox Flow Battery Prototype Development
[38] Clayton H, Makhubela B, Su H, Smith G and Moss J 2009 Synthesis, characterization, reactivity and molecular structure of arene-osmium complexes: A new synthetic entry into (η6-arene)osmium(II) chemistry *Polyhedron* **28** 1511–17
[39] Jia G 2007 Recent progress in the chemistry of osmium carbyne and metallabenzyne complexes *Chem. Rev.* **251** 2167–87

Acknowledgments

I would like to acknowledge Mintek for allowing me the opportunity to work on this project and for the funding of the entire project.

IOP Conf. Series: Materials Science and Engineering **839** (2020) 012015 doi:10.1088/1757-899X/839/1/012015

Technological, cycle and actual productivity of the surfacing process in the restoration of parts of forest machines

O Timokhova, O Burmistrova, D Shakirzyanov, V Burmistrov, D Snopok, R Timokhov, N Shol and E Mikhailenko

Ukhta State Technical University, 13, PervomaiskayaStreet, Ukhta, Komi Republic, 169300, Russia
chonochka@mail.ru

Abstract. Most forestry machines and mechanisms operate in adverse conditions: an abrasive medium at low temperatures, which impede lubrication of parts and components of mechanisms and affect the mechanical characteristics of metals; under significant loads caused by working conditions in the forest. Therefore, the issues of operation and repair of precisely logging machines are most relevant. Scientific research and best practices of repair enterprises currently have a significant number of studied methods for building up worn parts in the recovery process. Scientific research and best practices of repair enterprises currently have a significant number of studied methods for building up worn parts in the recovery process. The requirement to ensure high productivity of the extension operation is relevant and important. The most high-performance methods are metallization, casting, electro contact, surfacing. But the first two methods do not provide guaranteed adhesion to the base metal. When applying metal to a worn surface by surfacing, good adhesion and the possibility of obtaining a wear-resistant surface are ensured. When floating metal overhead, technological work is performed, which means "work to change the shape, properties, condition and position of the subject of labor. "Productivity of the equipment is determined by the amount of technological work performed for a certain period of time.

1. Introduction

One of the main tasks of the development of repair production is a constant and significant increase in the productivity of technological processes, including in restoration operations. Without considering productivity, it is impossible to correctly carry out a technical and economic assessment of the method of restoration. Labor productivity in the performed operations of the technological process cannot be considered according to the achieved values but should be evaluated by a system of indicators characterizing, in addition to the costs of living labor, the costs of social, past labor. The efficiency of the process with enough fullness can be determined by indicators: 1) equipment performance; 2) the complexity, energy intensity and metal consumption of the process. To determine the effect of various factors on productivity, consider technological, cyclic and actual performance.

2. Technological productivity

The productivity of the possible, clean operation of the equipment, or the productivity of continuous technological work, without any time expenditures for auxiliary operations, idling, preparatory and final operations.

In general

$$П_T = A \cdot V = K \cdot N_C \quad (1)$$

where A – load indicator; V – process speed indicator; K – process speed indicator; N_c – power consumption.
Cyclic performance is considered taking into account the time losses in the cycle:

$$П_Ц = П_T \cdot \eta_{t_Ц} \quad (2)$$

where $\eta_{t_Ц}$ – cycle time utilization.
Actual performance is considered taking into account all time losses, including off-cycle:

$$П_д = П_T \cdot \eta_t = \eta_t \cdot K \cdot N_C \quad (3)$$

where η_t - the total time utilization.
Surfacing technological productivity is estimated by the amount of deposited metal per unit time.
Can record:

$$П_T = F_{np} \cdot V_{np} \cdot \eta_n \quad (4)$$

or

$$П_T = 0,785 \cdot d_n^2 \cdot V_n \cdot \eta_n \quad (5)$$

where $\eta_n = 1 - \psi$.

In the expression (5), there are no electrical parameters, it is understood that the power source provides the melting of the entire amount of wire supplied (the performance of the metallization apparatus does not depend on the current, but on the wire feed speed and its diameter) [80].

Thus, the limiting values of technological productivity of the surfacing installation are limited:
1) The power of the power source;
2) The feed rate and the cross-section (diameter) of the electrode wire (Figure 1).

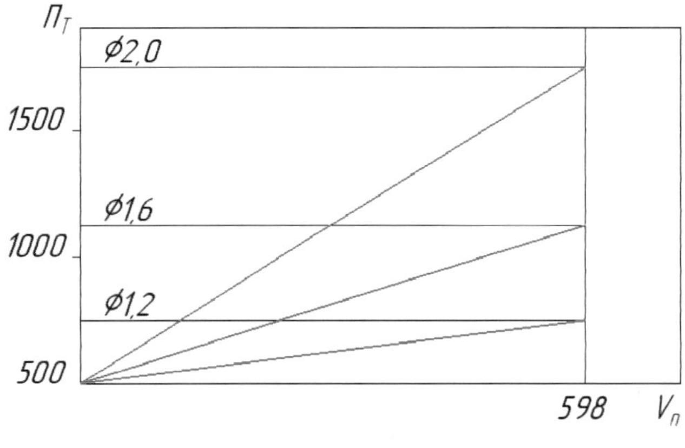

Figure 1. $П_T$ limitation by diameter and electrode wire feed speed.

3. Cycle performance

When surfacing a part, in addition to the main technological operation, it is necessary to perform a number of auxiliary operations that ensure the performance of the main technological work. The cycle time consists of the main technological time $t_T$of the auxiliary time $t_в$, and the coefficient of utilization of the cycle time

$$\eta_{t_ц} = \frac{t_T}{t_T + t_в} \tag{6}$$

The auxiliary time consists of: time for installation and removal of the part t_1, time for turning on and off the equipment t_2, time for clogging the oil channels of the part to protect against welding t_3, transition time t_4, time to clean equipment parts (for example, from metal spatter) t_5, time to control t_6. In determining $t_в$, it is necessary to consider the possible combination of time. To determine the technological time, the expression

$$t_T = \frac{V_Г}{K_v \cdot N_c} \tag{7}$$

Where $V_Г$ – is the volume of metal growth.

The volume of metal growth can be represented

$$V_Г = \pi \cdot d \cdot l(\delta_1 + f) \tag{8}$$

where d, l – dimensions of the deposited area; δ_1 – amount of wear; f – machining allowance.
Then

$$t_T = \frac{\pi \cdot d \cdot l(\delta_1 + f)}{K_v \cdot N_c} \tag{9}$$

To determine t_T, a nomogram is proposed as Figure 2 was used to determine t_T, the volume of metal growth.

To identify the nature of the change in the utilization of the cycle time, we accept: d = 30-110 mm; l = 50-300 mm; δ_1 = 1.0 mm; K_v = 9.7 cm²/MJ; η = 0.5; I_H= 150-400 A. The calculation results are presented in Table 1.

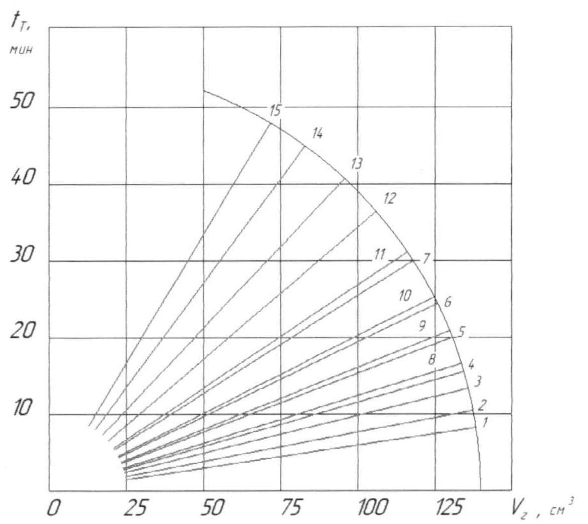

Figure 2. Nomogram for definition t_T. For d_{np}= 2.0: 1;2;3;4;5;6;7, respectively V_{np}= 2,3; 2.9; 3.8; 4.8; 6.0; 7.9; 11 cm. For d_{np}= 1.2: 8.9; 10; 11; 12; 13; 14; 15, respectively V_{np}= 2.3; 2.9; 3.8; 4.8; 6.0; 7.9; 11.0; 12.6 cm/s.

Table 1. The size of the deposited area both in diameter and in length.

	Plotsize ($d \times l$, cm²)	t_T	$t_в$	$t_ц$	$\eta_{t_ц}$
1	20	3.0	0.93	3.93	0.76
2	50	5.4	0.98	6.38	0.85
3	80	6.7	1.03	7.73	0.87
4	110	7.5	1.08	8.58	0.87
5	140	8.1	1.13	9.23	0.88
6	170	8.5	1.18	9.68	0.88
7	200	8.8	1.23	10.63	0.88
8	230	9.1	1.28	10.38	0.88
9	260	9.2	1.33	10.53	0.88
10	290	9.3	1.38	10.68	0.88

Table 1 shows that with an increase in the size of the deposited area both in diameter and in length with a simultaneous increase in arc power $\eta_{t_ц}$, when surfacing on small surfaces is 0.76-0.85. But later it stabilizes with value 0.85-0.9. At a constant arc power, which is typical for surfacing surfaces of the same diameter, but of different lengths, a change in a wider range should be expected, which is confirmed by the data in Table 2 (Accepted: d = 5.0 mm; l = 50-300 mm; I_H = 180 A; δ_1= 1 mm; f = 0.3-1.2 mm).

Table 2. The section length for each indicator.

Indicators	Section length (mm)										
	50	75	100	125	150	175	200	225	250	275	300
t_T **(min.)**	2.86	4.28	5.72	7.14	8.56	10.0	11.43	12.85	14.3	15.72	17.15
$t_в$ **(min.)**	0.93	0.98	1.03	1.08	1.13	1.18	1.23	1.28	1.33	1.38	1.43
$t_ц$ **(min.)**	3.79	5.26	6.75	8.22	9.69	11.18	11.18	14.13	15.63	17.10	18.58
$\eta_{t_ц}$	0.75	0.81	0.85	0.87	0.88	0.89	0.9	0.91	0.92	0.92	0.92

The utilization rate of the cycle time direct welding of the considered size intervals is 0.75-0.92; Cycle throughput will be calculated according to Equation 10.

$$П_Ц,= (0,75\text{-}0,92)\ П_T \tag{10}$$

4. Actual performance

When surfacing parts, the time is inevitable for preparatory and final operations, downtime and other time losses for various organizational and technical reasons, i.e. only an hour of shift work time is spent on the main and auxiliary operations of the work cycle - cycle work. The value of out-of-cycle time losses per shift can be characterized by the coefficient of time use as Equation 11.

$$\eta_t = \frac{\sum t_ц}{\sum t_ц + \sum t_n} \tag{11}$$

where$\sum t_ц$– cycle time per shift; $\sum t_n$– time spent on off-cycle losses.
When surfacing operation:

$$\sum t_n = T_{n_з} + T_{доп} \cdot n \qquad (12)$$

Where $T_{n_з}$ is the preparatory-final time for a batch of parts (it is assumed that a batch of parts is fused for a shift), consists of time for: receiving a task, a tool, preparing and cleaning a workplace, installing and removing tools and devices, setting up and adjusting equipment, commissioning finished products; $T_{доп}$ – time for maintenance of the working bridge and natural necessities, etc .; n – is the number of parts per shift.

Coefficient η_t can be represented in Equation 13.

$$\eta_t = \eta_{t_ц} \cdot \eta_{t_{n_з}} \cdot \eta_{t_д} \qquad (13)$$

where $\eta_{t_ц}$ – coefficient taking into account the loss of time on the preparatory-final operations; $\eta_{t_д}$ – coefficient taking into account the loss of time for organizational and technical reasons.

Otherwise

$$\eta_t = \frac{t_T \cdot n}{(t_T \cdot t_в \cdot t_д) \cdot n + T_{n_з}} \qquad (14)$$

Then the actual performance of the surfacing installation was confirmed using either one of the Equation as below:

$$П_д = K_v \cdot N \cdot \eta_{t_ц} \cdot \eta_{t_{n_з}} \cdot \eta_{t_д} \qquad (15)$$

or

$$П_д = K_v \cdot N \frac{n \cdot t_T}{n(t_T + t_в + t_д) + T_{n_з}} \qquad (16)$$

or

$$П_д = F_{np} \cdot V_{np} \cdot \eta_n \cdot \frac{n \cdot t_T}{n(t_T + t_в + t_д) + T_{n_з}} \qquad (17)$$

Actual productivity is directly proportional to the technological productivity of the equipment and the overall coefficient of use of time η_t, which increases with a decrease in the size of the batch of parts (per shift) and with a decrease in out-of-cycle. Performing actions by the operator on the surfacing installation combine the work of a locksmith and a welder. Combined work on the management of the machine and its maintenance: removal and installation of parts, cleaning parts of the equipment from splashes, setting up the machine for preset modes, etc.

We accept that $t_{обс}$ includes time elements on: care of equipment, working bridge; elimination of minor problems; regulation of the regime; turn on, turn off, power on the installation; layout and cleaning of tools; cleaning equipment from metal spatter.

5. Methodology of experimental-design studies of energy intensity

From the expression of the specific and hourly productivity (5) it can be seen that they depend on the amount of wire fed F_{np} V_{np} Coefficient of performance η installation; arc power $N_д$; spray coefficient ψ at the optimal values of the remaining parameters of the mode. At $F_{np} = Const$ according to the experimental I_H can be expressed through V_{np} by an equation of the type $y = k + bx$ ((under conditions of self-regulation of the arc). Coefficient of performance. η installation, summarizes Coefficient of performance of all hotel consumers of the installation, the greatest losses occur in the power source. Differentially experimentally investigated η and Coefficient of performance. installation η_u source in the download area 0,1-0,9 ($K_З=0,1\text{-}0,9$), which takes place during surfacing of restored parts.

Productivity was calculated as the amount of deposited metal to the total energy consumption of the installation. The coefficient ψ in the calculations is adopted (according to the results of experiments) 7%, the transfer coefficient is η_n is 0.93, respectively. For a comparative analysis, the performance is calculated and graphically shown for the converter and rectifier, with respect to only 1.6 and 2.0 mm

IOP Conf. Series: Materials Science and Engineering **839** (2020) 012015 doi:10.1088/1757-899X/839/1/012015

wires, when determining the graphical dependence of energy consumption, we considered differentially for technological, cyclic and actual performance. The utilization factors are times: cyclic and total - calculated on the basis of reference materials, as applied to the polygons of the distribution of parts by diameter and length.

Research results η and η_u в depending on K_3 (Figures 3, 4) for Power Supply 1 and Power Supply 2.

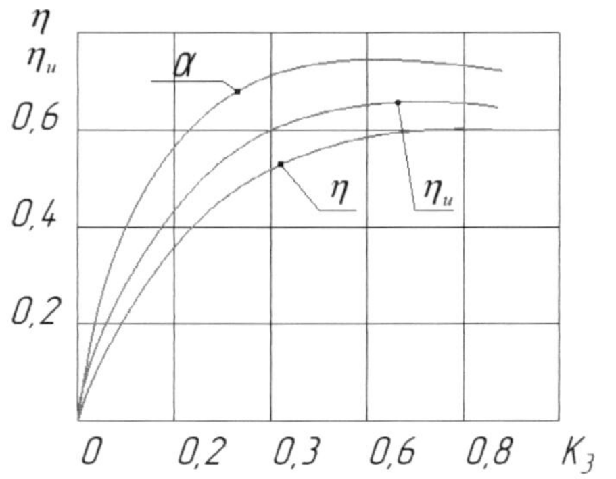

Figure 3. Dependence η, η_u on K_3 (Power Supply 1).

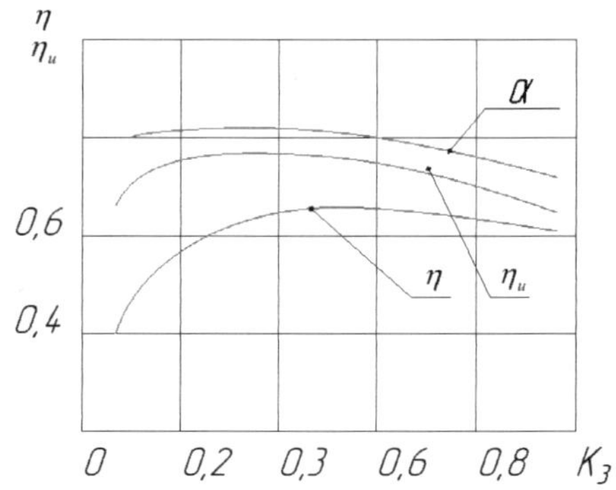

Figure 4. Dependence η, η_u on K_3 (Power Supply 2).

Given different values of I_H, , we calculate the specific productivity of surfacing installations from power supplies 1 and 2. The graphical dependence is presented in Figure 5.

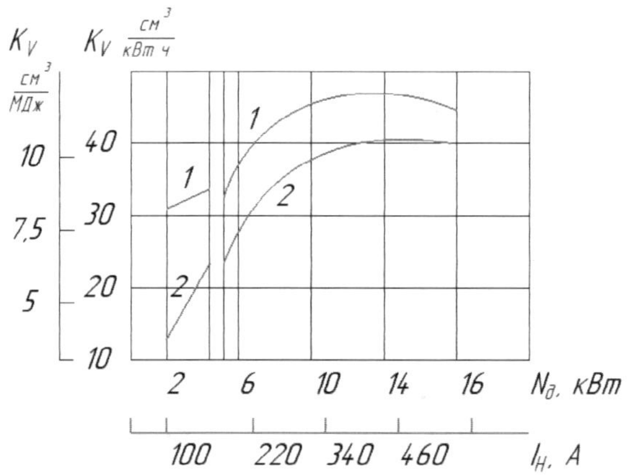

Figure 5. Actual installation performance:
1 - Power supply 1; 2 - Power Supply 2.

6. The results of experimental-design studies of energy intensity

The K_v indicator gives a specific assessment, characterizes the energy excellence of the installation at various load factors, indicates the rational area of use of the installation according to K_3. The $\Pi_Ц$ и $\Pi_д$ values are much smaller, which is explained by the low values of time coefficients. With regard to repair of parts, the ratio and nature of the change in the $(\eta_{t_ц}, \eta_t)$. With regard to repair of parts, the ratio and nature of the change in the $\Pi_Ц$ и $\Pi_д$ was shown in Figure 6.

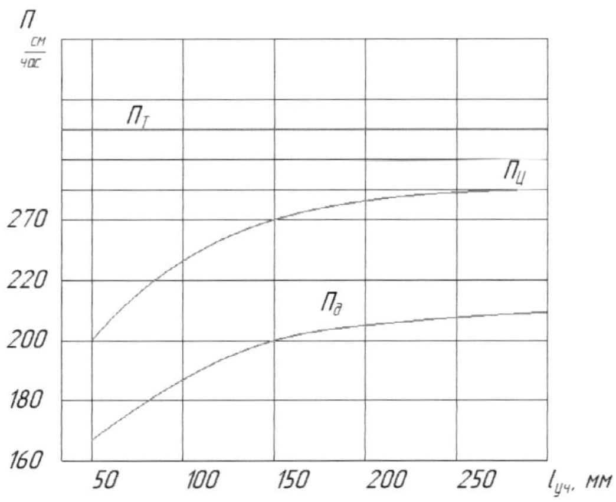

Figure 6. The nature of the change in $П_Ц$ and $П_о$ depending on l of the site.

Low Kv values at low $K_З$ load factors are explained by significant energy losses in the surfacing installation itself.

7. Conclusions

1. The expressions obtained show that the work of applying a metal coating to a unit of consumed energy is determined by: 1) the cross section of the electrode wire and its feed rate; 2) the efficiency of the installation and use of metal; 3) the energy consumption of the arc. You can increase the value of Kv by increasing; 1) wire cross-section; 2) wire feed speed; 3) set the efficiency. Increasing the value of the arc power during $F_{np} \cdot V_{gh} = Const$ will reduce the specific productivity of the installation. It is necessary to use the minimum arc power, which provides satisfactory melting of the feed wire.
2. The technological productivity of the surfacing installation is determined by the amount of wire that can be supplied and is limited by the power of the power source.
3. By reducing the amplitudes of fluctuations in the electrical parameters during periods of microcycles, it is possible to increase $П_Т$ by 5-10%.
4. The actual productivity is 30-50% of $П_Т$ and can be improved by reducing all kinds of time losses.
5. The main factors of actual performance is: time; specific productivity; power.

8. References

[1] Timokhova O, G Koptyaeva and N Shol 2011 Increase of corrosion resistance of technological equipment and machine parts *Scientific and Technical Bulletin of the Volga Region* **3** 141-145
[2] Fisher R 1951 *The Design of Experiments 6-th ed* (London: Oliver and Boyd)
[3] Box GEP and Wilson KB 1951 On the Experimental Attainment of Optimum Conditions *J.Roy. Statist.Soc.* **13** 1-45
[4] Bowden Y and Kelly P 1967 The cristallography of the pressure induced phase transformation in iron alloys *Acta Met.* **15** 1489-1500
[5] Timokhova O and Burmistrova O 2018 Analysis of the influence of surfacing modes on the structure of metal *Natural and technical sciences* **11** 313-317
[6] Timokhova O 2017 Investigation of the influence of technological surfacing regimes on the physical and mechanical properties of the coating *News of the Samara Scientific Center of the Russian Academy of Sciences* **9**
[7] Timokhova O, Burmistrova O, Sirina E and Timokhov R 2017 Effect of plasma spraying modes on material properties of internal combustion engine cylinder liners *Int. Conf. on mechanical engineering, automation and control systems 2017 - simulation and automation of production engineering (Tomsk)* vol 327 (*Bristol: IOP Publishing Ltd.*) p 022101
[8] Timokhova O, Burmistrova O and Sirina E 2017 Optimization of cylinder liner plasma spraying

mode *Int. Conf. on innovations and prospects of development of mining machinery and electrical engineering 2017 (Saint-Petersburg)* vol 87 (*Bristol: IOP Publishing Ltd.)* p 082049

Properties of fine Sr^{2+}-substituted hydroxyapatite synthesized using ultrasonication

H Minamisawa[1]*, K Nomura[1], A Onda[2], T Umegaki[1] and Y Kojima[1]

[1]Department of Materials and Applied Chemistry, Faculty of Science and Engineering, Nihon University, 1-8, Kanda-Surugadai, Chiyoda-ku, Tokyo 101-8308, Japan
[2]Department of Chemistry and Biotechnology, Faculty of Science and Technology, Kochi University, 2-5-1, Akebono-cho, Kochi-city, Kochi, 780-8520, Japan

Abstract. This study investigated the properties of fine Sr^{2+}-substituted hydroxyapatite (SrHAp) synthesized using ultrasonication. SrHAp was synthesized by the addition of phosphoric acid solution to an ultrasonicated calcium hydroxide suspension containing strontium hydroxide. The X-ray diffraction peaks for SrHAp were shifted to low angle with increasing Sr/(Ca + Sr) atomic ratio. X-ray diffraction, energy dispersive X-ray spectroscopy and inductively coupled plasma spectroscopy measurements confirmed that HAp and strontium apatite ($Sr_{10}(PO_4)_6(OH)_2$;SrAp) formed a complete solid solution. When the Sr/(Ca + Sr) atomic ratio was 0.3 or less, the (Ca + Sr)/P atomic ratio was 1.67 or less, and the specific surface area was also large. The bending strength of a SrHAp sintered body was increased and the bending strength of the SrAp sintered body became 40 MPa when the Sr/(Ca + Sr) atomic ratio was above 0.3. The ethanol conversion rate using SrAp as catalyst was 7% for 1 hour and was 40% of HAp reagent. However, selectivity of SrAp from ethanol to butanol was 70%.

1. Introduction

Apatite is represented by the general formula $M_{10}(ZO_4)_6X_2$, where the M^{2+} ions are generally Ca^{2+}, Sr^{2+}, Cd^{2+}, or Pb^{2+}, ZO_4^{3-} represents PO_4^{3-}, or VO_4^{3-} and X^- is OH^-, F^-, or Cl^- [1]. Hydroxyapatite (HAp) has a typical apatite structure and is represented by $Ca_{10}(PO_4)_6(OH)_2$. HAp can be used as a biomaterial substitute for teeth and bones, an ion-exchanger for repair heavy metals, a chromatographic filler for absorption of proteins and a catalyst for alcohol conversion [2-11].

HAp synthesis methods include dry and wet method, the microwave method and the ball-milling method [12-18]. We have been performing a series of studies on the synthesis of fine calcium compounds using ultrasonication. For example, calcium carbonate with specific surface area of 100 m^2/g was easily formed by direct ultrasonication [19, 20]. HAp with a specific surface area of 200 m^2/g or more was synthesized by reacting calcium hydroxide and phosphoric acid while irradiating with ultrasound [21].

Strontium-substituted hydroxyapatite (SrHAp) in which a portion of the Ca^{2+} ions are replaced with Sr^{2+} ions, is considered to have catalytic and ion-exchange performance similar to HAp. Furthermore, SrHAp is thought to be effective for improving the sinterability of HAp. SrHAp has been synthesized by urea hydrolysis, hydrothermal treatment, and addition of hydrogen phosphate at high temperature [22-30]. SrHAp has often been synthesized by addition of phosphate to nitrate containing Ca^{2+} and Sr^{2+} ions, and it was reported that SrHAp was produced directly, although the reaction process was unclear. Fine Sr-HAp particles were not obtained because it was slowly formed at high temperature.

Onda et al. have conducted a series of studies on the conversion of alcohol using HAp, and examining the catalytic ability of apatite structure such as HAp and strontium apatite ($Sr_{10}(PO_4)_6(OH)_2$, SrAp) [9, 10]. It is expected that the catalytic ability could be improved by increasing the specific surface area. The purpose of the present study was to synthesize fine SrHAp and SrAp particles using ultrasonication, and investigate the relationship between the amount of Sr^{2+} solid solution and the specific surface area, and the conditions for SrAp synthesis. The catalytic activity of synthesized HAp, SrHAp and SrAp was also investigated.

2. Materials and methods

2.1. Materials

Strontium hydroxide octahydrate, calcium hydroxide and phosphoric acid were used for the synthesis of SrHAp and SrAp. Strontium hydroxide octahydrate and phosphoric acid were special grade reagents obtained from Kanto Chemical, and calcium hydroxide (99.99%) was manufactured by Shiraishi Calcium Co., Ltd.

2.2. Experiment

SrHAp was synthesized by combining strontium hydroxide octahydrate and calcium hydroxide at a Sr / (Ca + Sr) atomic ratios in the range of 0.1 to 0.9 and suspending them in 60 cm^3 of pure water in a stainless beaker. The suspension was placed in a water tank and irradiated with ultrasound from the outside to make the suspension temperature 30 ℃. This ultrasonication device (BJ-2566ST and S8580LP-5) was used at a frequency of 80 kHz with a power at 200 W, and five irradiation directions were used. 60 cm^3 of 0.1 mol/dm^3 phosphoric acid aqueous solution was then quickly added to the suspension during ultrasonication. The (Ca + Sr) / P atomic ratio was 1.67. The maximum reaction time was 10 min. The product was centrifuged at 10,000 rpm for 5 min and the precipitate was filtered using a membrane filter. The product was then kept in a vacuum dryer at 70 ℃.

The synthesis of SrAp was conducted by reacting strontium hydroxide octahydrate in water and phosphoric acid aqueous solution with ultrasonication in the same way as the SrHAp synthesis. After the reaction, the resulting suspension was filtered, washed with pure water, and then dried under vacuum. SrHAp sintered body was obtained by molding SrHAp powder at 100 MPa and firing at 1000 °C for 3 h. The catalytic properties of SrHAp were studied by adding ethanol at 350 °C. The obtained sample was characterized using powder X-ray diffraction (XRD; Rigaku MultiFlex, Japan), infrared (IR) spectrometer (JASCO FT/IR-4600, Japan), field-emission electron microscopy (FE-SEM; Hitachi D-4500, Japan) and scanning transmission electron microscopy (STEM, JEOL JEM-F200). XRD was conducted at a scan rate of 8°/min, and IR absorption measurements were performed using the KBr disk method. SrHAp was dissolved using 3 mol/dm^3 HCl solution, and Sr^{2+} and Ca^{2+} in HAp particle were quantified using ICP. The catalytic activities of SrHAp and SrAp were examined for the conversion of ethanol to butanol at 350 °C.

3. Results and discussion

3.1. Synthesis of strontium-substituted hydroxyapatite

The specific surface area of HAp synthesized by the addition of phosphoric acid solution to a high-purity calcium hydroxide suspension with ultrasonication was 263 m^2/g. Since the radius of Ca^{2+} ions is 0.099 nm and that of Sr^{2+} ions is 0.112 nm, diffraction peaks of SrHAp shifted to low angle side when Sr^{2+} ions incorporated in HAp structure.

Figure 1 shows XRD patterns for HAp and SrHAp synthesized at different Sr/(Ca+Sr) atomic ratio. The diffraction peaks for HAp with specific surface area of 263 m^2/g increased very low crystallinity. The (211) diffraction peak for HAp near $2\theta = 32°$ shifted continuously to low angle with increasing Sr/(Ca+Sr) atomic ratio, and the crystallinity gradually increased. The diffraction peaks for SrAp without Ca^{2+} ions indicated higher crystallinity than that for HAp. Figure 2 shows enlarged XRD patterns around 2θ = 29 – 33° to clarify the peak shift. It was confirmed that diffraction peak continuously shifted to a low angle of ca. 2θ = 1.5° with the change in the composition from HAp to SrAP. Both HAp and SrAp crystal system are hexagonal. When the crystal system is the same and the difference in cation radius is small, a solid solution is formed. For this reason, it is assumed that HAp and SrAp form a solid solution.

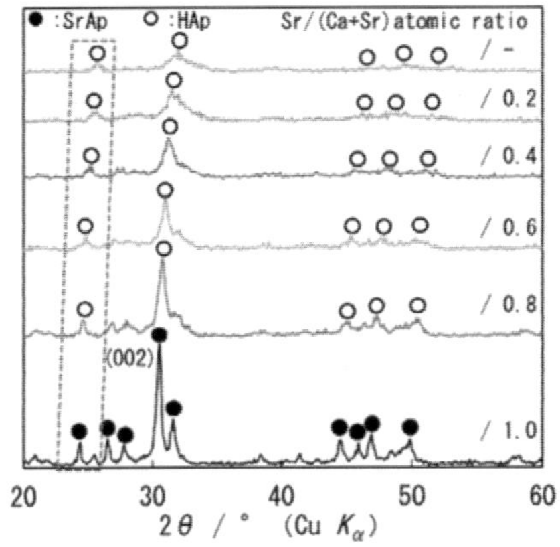

Figure 1. XRD patterns for Hap and SrHAp synthesized at different S/(Ca+Sr) atomic ratio.

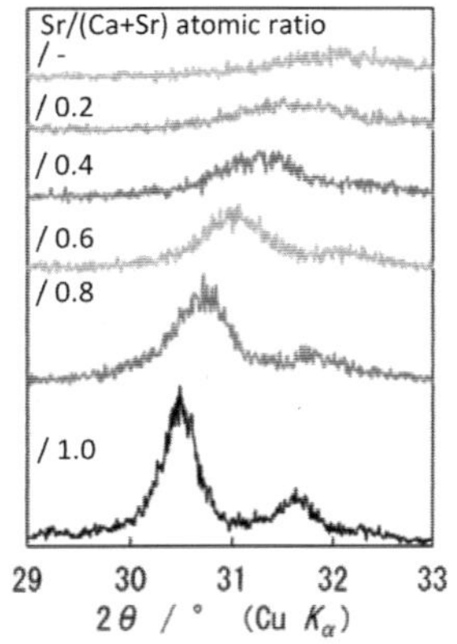

Figure 2. Enlarged XRD of Figure 1.

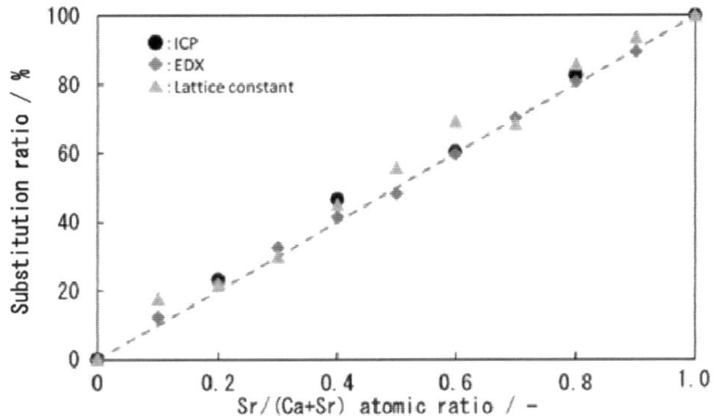

Figure 3. Effect of the Sr/(Ca+Sr) atomic ratio on substitution ratio of Sr^{2+} ion in Hap structure.

The substitution ratio was calculated from the movement of the diffraction peak due to the (002) plane of HAp near 2θ = 26° and was also obtained from EDX and ICP. Figure 3 shows the effect of the Sr/(Ca+Sr) atomic ratio on substitution ratio of Sr^{2+} ion in HAp structure determined from X-ray diffraction peak, EDX and ICP measurements. The broken line shows the relationship between the Sr/(Ca+Sr) atomic ratio and the substitution ratio. If a point is on the line, it indicates that the substitution was successfully achieved according to the initial Sr/(Ca+Sr) atomic ratio. The substitution ratio of Sr^{2+} ion in HAp determined from XRD was the same or slightly higher than this broken line, but was almost on the line. These results confirmed that HAp with any Sr^{2+} content can be synthesized by changing the Sr/(Ca+Sr) atomic ratio. The ICP result also confirmed that substitute ratio increased with the Sr/(Ca+Sr) atomic ratio. The same results were obtained using EDX.

3.2. Properties of SrHAp and SrAp

The properties of synthesized SrHAp, such as the specific surface area, (Ca+Sr)/P atomic ratio, sinterability and catalyst acitivity, were investigated. Figure 4 shows the effect of the Sr/(Sr+Ca) atomic ratio on specific surface area of SrHAp. The specific surface area of HAp was 263 m^2/g. However, the specific surface area rapidly decreased with increasing Sr/(Ca+Sr) atomic ratio. The specific surface

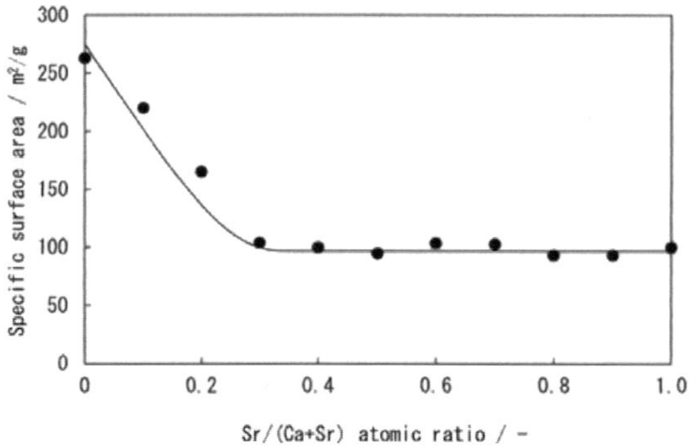

Figure 4. Effect of the Sr/(Ca+Sr) atomic ratio on specific surface area of SrHAp.

area for a Sr/(Ca+Sr) atomic ratio of 0.3 was approximately 100 m^2/g, and no change was observed thereafter. XRD measurements indicated the crystallinity of SrHAp was increased by the addition of Sr^{2+} ions. The crystallinity of SrHAp gradually improved as the Sr^{2+}-substitution ratio was increased. Although the difference in the ionic radius between Sr^{2+} and Ca^{2+} ions is small, it was confirmed that SrHAp particles grow rapidly when Sr^{2+} ions were substituted into the HAp structure. The densities of HAp and SrAp were 3.16 and 4.08 g/cm^3, respectively. Therefore, even when the particle size is the same, the specific surface area decreased when Sr^{2+} ions was replaced by HAp. However, due to the large decrease in specific surface area, it is considered that the particles produced with the substitution of Sr^{2+} ions became larger. The solubility of SrAp is greater than that of HAp; therefore, it was expected that SrAp with a small specific surface area would be generated. However, the specific surface area of the SrAp formed was 100 m^2/g. Such fine SrAp has never been obtained. The particle size can be determined by the following equation (1).

$$\text{Particle size} = \frac{6}{(\text{specific surface area} \times \text{density})} \tag{1}$$

The SrAp particle size obtained from the calculation was 15 nm. Although this SrAp was ultrasonicated by the indirect method, SrAp with the same specific surface area was obtained by the direct method; therefore, the particles are smaller because the specific surface area is constant despite containing a large amount of Sr^{2+} ions.

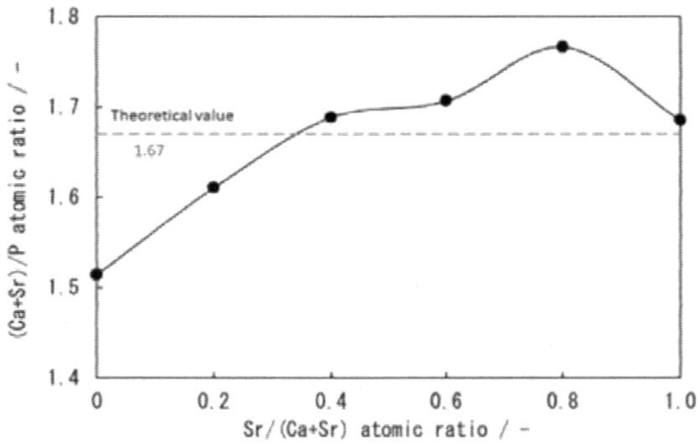

Figure 5. Effect of the Sr/(Ca+Sr) atomic ratio on (Ca+Sr)/P atomic ratio of SrHAp.

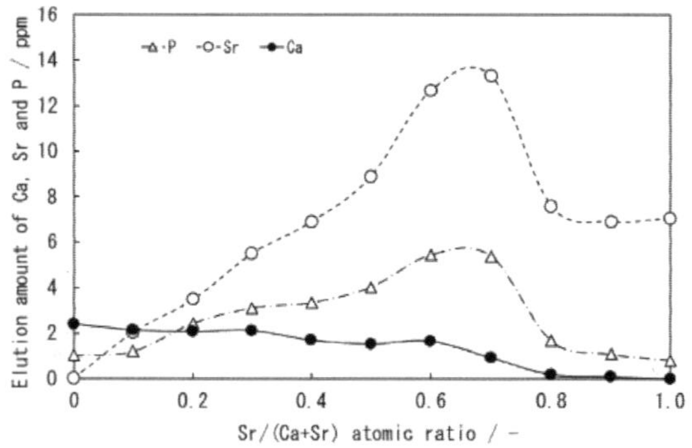

Figure 6. Elution amount of Ca, Sr and P from SrHAp.

Figure 5 shows the effect of the Sr/(Ca+Sr) atomic ratio on the (Ca+Sr)/P atomic ratio for the products. The (Ca+Sr)/P atomic ratio for HAp was 1.52, which was smaller than the theoretical value of 1.67, i.e, the synthesized HAp was Ca defect HAp. The (Ca+Sr)/P atomic ratio increased with increasing Sr/(Ca+Sr) atomic ratio and was 1.68 when sr/(Ca+Sr) atomic ratio was 0.4. Considering the specific surface area results, the specific surface area of the particles with defects was large, which suggested that stable particles were formed when the (Ca+Sr)/P atomic ratio approached the theoretic atomic ratio.

Figure 6 shows the result of a SrHAp powder dissolution. Ca^{2+}, Sr^{2+} and PO_4^{3-} ions were measured using ICP spectroscopy. The eluate of HAp contained Sr^{2+} ions, which was derived from the $Ca(OH)_2$ raw material. The amount of Ca^{2+} ions in the eluate decreased linearly until a Sr/(Ca+Sr) atomic ratio of 0.7, whereas Sr^{2+} and PO_4^{3-} ions increased linearly. The amount of each ion eluted decreased when the Sr/(Ca+Sr) atomic ratio was 0.7 or more. However, in the case of a Sr/(Ca+Sr) atomic ratio of 0.8, Sr^{2+} and PO_4^{3-} ions were dissolved rapidly, while the Ca^{2+} ions were almost insoluble. Figure 6(b) shows the change in the (Ca+Sr)/P atomic ratio at this time; the (Ca+Sr) atomic ratio was 1 when the Sr/(Ca+Sr) atomic ratio was in the range from 0.2 to 0.7.

Samples synthesized at each Sr/(Ca+Sr) atomic ratio were uniaxially pressed and then fired at 1000 ℃ to produce a sintered body for bending strength tests.

Figure 7 shows the effect of Sr/(Ca+Sr) atomic ratio on the bending strength of sintered bodies produced for each Sr/(Ca+Sr) atomic ratio. The bending strength of the HAp sintered body was 15 MPa. However, the bending strength decreased up until a Sr/(Ca+Sr) atomic ratio 0f 0.3. This was also related to the decrease in the specific surface area down to a Sr/(Ca+Sr) atomic ratio of 0.3, and the bending strength decreased as the SrHAp particle size increased. Above a Sr/(Ca+Sr) atomic ratio of 0.3, the bending strength increased linearly, and the bending strength of SrAp was 40MPa, which was more than twice that of HAp. HAp particles containing Sr^{2+} ions are considered to have improved strength due to easy sintering.

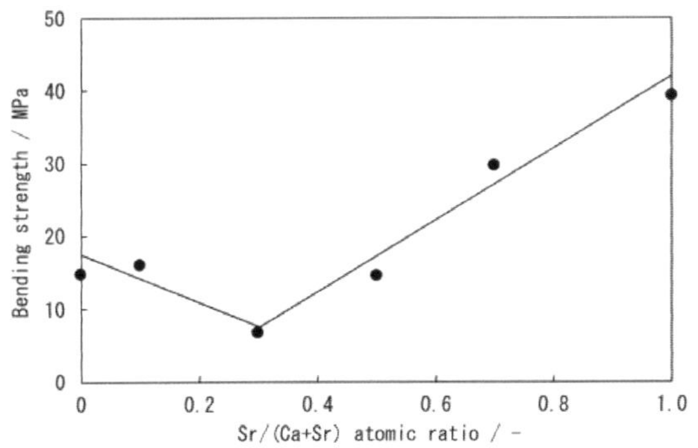

Figure 7. Effect of the Sr/(Ca+Sr) atomic ratio on bending strength of SrHAp.

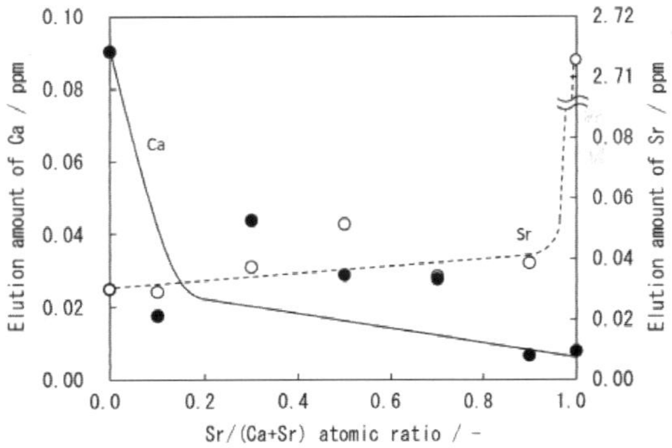

Figure 8. Elution amount of Ca and Sr from SrHAp sintered body immersed in water.

Figure 8 shows the solubility of each SrHAp sintered body in water. The amount of dissolved Ca^{2+} ions was 0.09 ppm and a trace amount of Sr^{2+} was detected. The existence of Sr^{2+} ions depended on the raw material $Ca(OH)_2$. The amount of dissolved cations from the SrHAp sintered body was significantly reduced. By increasing the Sr/(Ca+Sr) atomic ratio of the sintered body, the amount of dissolved Ca^{2+} ions was reduced, while the amount of dissolved Sr^{2+} ions was slightly increased. The reason for the decrease in the amount of cations dissolved is presumed to be that structural distortion was eliminated by heating. It was also possible that heating caused grain growth, which reduced the amount of dissolution. The amount of Sr^{2+} ions dissolved from the SrAp sintered body was 2.71 ppm. It was expected This is because SrAp shows high dissolution amount. It was expected that finer powders would be easier to sinter, but in fact SrAp powders with larger particles were easier to sinter and the strength of the sintered body increased.

Onda et.al. investigated alcohol conversion using SrHAp or SaAp as a catalyst. The conversion from ethanol to butanol can be performed in two stages, but it can be performed in one stage using a catalyst such as HAp. Butanol has a 30 % higher calorific value than ethanol, and it can be used by mixing with gasoline.

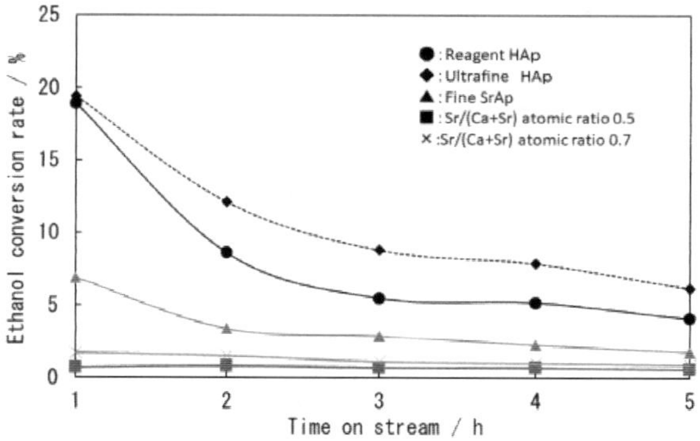

Figure 9. Effect of the Sr/(Ca+Sr) atomic ratio on the ethanol conversion rate over SrHAp or SrAp.

Figure 9 shows the effect of the Sr/(Ca+Sr) atomic ratio on the ethanol conversion rate over SrHAp or SrAp. The best conversion was obtained with ultrafine HAp. The total amount of ultrafine HAp was 1.3 times that of HAp reagent. Below that, the order of ethanol conversion rate was SrAp, SrHAp with a Sr/(Ca+Sr) atomic ratio 0.7, and SrHAp with a Sr/(Ca+Sr) atomic ratio of 0.5. The ethanol conversion rate over SrAp was 32% that of ultrafine HAp. When SrAp with specific surface area of 23 m^2/g, ethanol conversion rate was 11.3% as reported by Onda et.al.[9]. An Increase in the specific surface area of SrAp increased the ethanol conversion rate by 1.5 times. In addition, the selectivity toward butanol for HAp, SrHAp and SrAp were ca. 70%. The ethanol conversion rate could be increased by an increase in the specific surface area of HAp. Furthermore, It was thought that the reaction at a temperature lower than 350 ℃ will make the ethanol conversion rate of SrAp comparable to that of HAp reagent.

4. Conclusion

The aim of this study was to synthesize fine SrHAp and SrAp particles using ultrasonication and to examine their sintering and catalytic properties. XRD, EDX and ICP measurements indicated that HAp and SrAp formed a complete solid solution. Therefore, properties such as the specific were measured using SrHAp. The specific surface area of ultrafine HAp was 263 m^2/g, and the specific surface area of SrHAp in the Sr/(Ca + Sr) atomic ratio range of 0.3 to 0.9 and that of SrAp was ca. 100 m^2/g. The (Ca + Sr)/P atomic ratio for SrHAp in this range was 1.68, which was the same as the theoretical value. The (Ca + Sr)/ P atomic ratio for SrHAp with a Sr/(Ca + Sr) atomic ratio of 0.4 – 1.0 was 1.7. Fine SrAp sinterability was better than that for HAp sintered body. The bending strength of the SrAp sintered body was 40 MPa. When the specific surface area of the HAp or SrAp catalysts was large, the ethanol conversion rate over ultrafine HAp and fine SrAp has improved.

5. References

[1] Bigi A,Boanini E, Capuccini C and Gazzano M 2007 *Inorg.Chim.Acta* **360** 1009
[2] Akamatsu N 1993 *Asian.Med.J.* **36** 621
[3] Hench L 1998 *J.Am.Ceram.Soc.* **81** 1705
[4] Guo H, Xu C J Z, Luo P, Fu Z and Zhang J 2019 *Mater.Letter* **241** 176
[5] Jing N, Zhou A and Xu Q 2018 *J.Hazardous Mater.* **353** 89

[6] Wang Y, Liu Y,Lu H, Yang R and Yang S 2018 *J.Solid State Chem.* **261** 53
[7] Wang K, Wang M, Wang Q, Lu X, Zhang X 2017 *J.Euro.Ceram.Soc.* **37** 2509
[8] Yin G, Liu Z, Zhan J, Ding F and Yuan N, 2002 *Chem.Engineer.J* **87** 181
[9] Ogo S, Onda A and Yanagisawa K 2011 *Applied Cataly.A:General* **402** Issues1-2 188
[10] Ogo S, Onda A, Iwasa Y, Hara K and Yanagisawa K 2012 *J.Cataly.* **296** 24
[11] Mochales C, Wilson R M, Dowker S E P and Ginebra M P 2011 *J.Alloys Compounds* **609** 7389
[12] Toriyama M, Kawamura S, Ito Y and Nagae H 1989 *J.Ceram.Soc.Japan* **97** 554
[13] Chen R and Shen J 2020 *Mater.Letters* **259** 12688
[14] Silva C C, Grac M P F, Valente M A and Sombra A S B 2006 *J.Non-Crystall.Solids* **352** 1490
[15] Monma H and Kanazawa T 1972 *Nikka* **1972** 339
[16] Masuda Y, Matsubara K and Sakka S 1990 *J.Ceram.Soc.,Japan* **98** 1233
[17] Suwa K, Banno H, Mizuno M and Saito H 1993 *J.Ceram.Soc.Japan* **101** 659
[18] Bucur A I, Bucur R A, Szabadai Z, Mosoarca C and Linul P A 2017 *Mater.Character.* **132** 76
[19] Kojima Y, Yamaguchi K and Nishimiya N 2010 *Ultrasonics Sonochem.* **17** 617
[20] Kojima Y, Yamaguchi K and Nishimiya N 2009 *J.Soc.Inorg.Mater. Japan* **16** 171
[21] Kojima Y, Kitazawa K and Nishimiya N 2012 *Journal of Physics: Conference Series* **339** 012001
[22] Edwin N and Wilson P *Ultrasonic Sonochem.* **188-199** 50
[23] Carvalho E V, Paula D M, Andrade D M, Costa L S, Dias D F, Feitosa V P and Fechine P B A 2020 *J.Mechni Behavor Biomedi.Mater.* **103447** 101
[24] Cheng G, Zhang Y, Yin H, Ruan Y, Sun Y and Lin K 2019 *Ceram.Inter.* **45** 11073
[25] Suganthi R V, Elayaraja K, Hoshy M I A, Chandra V S, Girija E K and Kalkura S N 2011 *Mater.Sci.Engin.C* **31** 593
[26] Kaygili O, Keser S, Kom M, Eroksuz Y, Dorozhkin S V, Ates T, Ozercan I H, Tatar C and Yakuphanoglu F 2015 *Mater.Sci.Engin.C* **55** 538
[27] Abert J, Bergmann C and Fischer H 2014 *Crem.Inter.* **40** 9195
[28] Kavitha K, Subramanian R, Narayanan R and Udhayabanu V 2014 *Powder Tech.* **253** 129
[29] Pan H B, Li Z Y, Lam W M, Wong J C, Darvell B W, Luk K D K and Lu W W 2009 *Acta Biomater.* **5** 1678
[30] Urugan A B, Balossier G, Torres P, Michel J and Ferreira J M F 2009 *Mater.Sci.Engin.C* **29** 1006

IOP Conf. Series: Materials Science and Engineering **839** (2020) 012017 doi:10.1088/1757-899X/839/1/012017

Influence of organoclay on behaviour of polylactic acid (PLA)/polypropylene (PP) blend

Noor Hafizah Abdul Halim, Engku Zaharah Engku Zawawi, Ahmad Zafir Romli, Rozana Mohd Dahan, Nor Yuliana Yuhana

Faculty of Applied Science, Universiti Teknologi MARA, 40450 Shah Alam, Selangor, Malaysia

Abstract. In this study, thermoplastic composite by blending of Polylactic acid (PLA) and Polypropylene (PP) was prepared with fixed ratio of 75:25 % by weight. Varying content of organoclay were studied with different loading of 1,3,5 and 7 wt.% in order to study the potential of this nanofiller in the composite. The composites were prepared by using twin screw extruder. Characterizing of PLA/PP-Clay (PLA/PP-C) composites were analyzed by focusing on physical, thermal and morphological properties. Water absorptions was analyze to observe the percentage of water uptake when percentage loading of organoclay added were increased. Differential scanning calorimetry (DSC) was performed to analyze the thermal stability and thermal sensitivity of the composite. Meanwhile, x-ray diffraction (XRD) were observed to study the fracture surfaces of PLA/PP-Clay composites. In water absoptions analysis it was found that the percentage of water uptake were decreased when organoclay added at 7 wt.%. DSC results showed that melting temperature (Tm) were decreased when the loading of organoclay increased and the thermal stability was improved by addition of organoclay. Morphological analysis by XRD showed that exfoliation and intercalation happened in the system. It is suggested that these combination of PLA/PP-Clay have a good potential for the fabrication of nanocomposites.

1. Introduction

Polymer nanocomposites was investigated intensively by using nanoparticles which is only small loading of silicates needed but can exhibit better properties in nanocomposites fabrication. Development of composites by using microfiller have reached their limitation in optimization. The previous study found that their performance in strength, resistance to corrosion and thermal stability by using microfiller has a lot of trade off [1]. Nanocomposites was discovered in 1990s where Toyota was applied these polymer nanocomposites in the parts of their car's brand then these application was widely used during 2000s and above. Even though the era of nanocomposites was started to be used a long time ago, but the dispersion state of nanoparticles at low filler loading remain being challenges for the researchers to study their full potential of performance [2]. The performance of nanocomposites depends on their features such as their size, aspect ratio, surface area, volume fraction and also their compatibility with the other types of matrix. Nanocomposites was extensively investigate to develop enhancement in mechanical properties, improvement in thermal stability and sensitivity, gas barrier performance and reduce flammability to be used in variation of applications at low filler loading. The types of clay filler play the important role in order to develop better properties.

Montmorillonite clay (MMT) was one of the organically silicate that commonly used to their study of effects on the plastic blends to fabricate nanocomposite. Clay minerals can be categorized by 4 different types which is kaolinite group, smectite group, illite group and also chlorite group [3]. MMT was under smectite group which is its particle in sheets formed and has layers with a layer thickness about 0.96nm and the lateral dimension of 100-200nm. The layers divided into two types which are octahedral and tetrahedral structural of sheets which is content of aluminium or magnesium octahedral sheets between two silica-oxygen tetrahedral sheet. MMT is low cost and readily available. MMT was studied by other researchers in fabrication of MMT with other polymeric materials such as Poly(vinyl) alcohol (PVA), Polyethylene (PE), Polycaprolactone (PCL) and also Polypropylene (PP).

Polylactic acid (PLA) was gained attraction from researchers to investigate their functions and properties to produce biodegradable plastics composite in order to reduce the usage of conventional plastic sources from fossil-based polymers materials. PLA was identified as biogradable material which is it was derived from renewable sources such as potato, yarn, and sugar cane. PLA has good performance in mechanical strength, thermal stability and also eco-friendly material. But PLA also has high brittleness and stiff polymer with low elongation[4]. Because of this disadvantages in PLA properties, it was required to combine with the other type of polymer material in order ro develop it properties. For example blendinf of PLA with polypropylene (PP), High density polyethylene (HDPE) and PCL. Polypropylene (PP) is synthetic plastics material which is synthesized from petroleum-based sources. This material was widely used in the industry applications due to their flexible properties and also one of fastest growing types of thermoplastics. Based on the Department of Statistics in 2012, about 19,000 metric tons of waste from solid waste and plastic waste were produced from all the households in Malaysia [5]. The increment of plastics waste at the landfill was critical and most of plastics were dumped are from petroleum-based materials and eventually end up as nondegradable waste at the end. Therefore, the using of biopolymer materials such as PLA in product applicatrion is among the solution to reduce these problems.

Blending of plastics was gained attention from researchers since it can enhance the properties of composites and can reduce the cost of production. PLA is a hydrophilic aliphatic polyester material meanwhile PP is hydrophobic material behaviour. The blending of both materials will tend to cause of immiscibility between these two materials and the blending of matrix with filler also exhibit the problem. Afterwards it will cause low adhesion between materials and contribute to the low mechanical properties of the composite perfromance. In this case, compatibilizer was important to overcome that problem. Compatibilizer was functional as an aid during the processing then can exhibit development of adhesion bonding between all materials during blending process. Among of the compatibilizer that usually used in this blending process was maleic anhydride grafted polypropylene (MAgPP). The objective of this study to prepared and study the effect of organoclay filler on the thermal and morphological properties of PLA/PP blend.

2. Materials and method

2.1. Materials

PLA resin grade 2003D was purchased from Nature Works LLC with specific gravity of 1.24 g/cc, MFR 6.0 g/10 min and have transparent appearance. PP with melt flow index of 2.1 g/10 min with density of 0.9 g/cm^3 was produced by Lotte Chemical Titan Holding Sdn Bhd. Organoclay (Nanomer® I.44P) and maleic anhydride grafted polypropylene (MAgPP) were purchased from Aldrich.

2.2. Preparation of PLA/PP nanocomposite

PLA and PP blend with organoclay was melt blended in twin screw extruder TSE 16TC with fix ratio of 75 wt.% PLA and 25 wt.% PP. The loading of organoclay was varied at 1,3,5 and 7 wt.% and namely as PLA/PP-C1 until PLA/PP-C7. All materials were predried in oven at temperature 80°C for

8 hours prior of blending. Afterwards, all materials were premixing at the outside then blend in the twin screw extruder. The extrudate produced were crushed by using crusher in order to produce granulate shaped. Then, all materials were prepared according to the testing required.

2.3. Tesing and characterization

Water absorption analysis were done to observe the water uptake of the nanocomposites when organoclay filler were added in the PLA/PP blends. All samples were cut and dried in oven for 24 hours at 70°C. The samples were weighed by using electronic weight balance to record the dry weight of samples. Then, all samples were submerged in distilled water and immersed for 24 hours. The samples were taken out for weight measurement and wiped with tissue to remove water on the sample's surface. The weight gain percentage were calculated using equation 1.

$$\text{Water uptake \%} = \frac{wet\ weight - dry\ weight}{Dry\ weight} \times 100 \qquad (1)$$

For thermal analysis, differential scanning calorimetry (DSC) analysis were conducted by using Netzsch DSC 200 F3 to study the thermal behaviour of the composites. The samples weight of 8-10mg were heated from 30°C to 250°C under nitrogen atmosphere with the heating rate of 10°C/min and then held at this temperature for 5 min before proceed to second heating process in order to remove the thermal history and to ensure the materials melted uniformly. Then the samples heated again to 250°C at the same heating rate under nitrogen atmosphere for the second heating process. The glass transition temperature (Tg), crystallization temperature (Tc) and melting temperature (Tm) were determined and recorded at the heating curves.

Meanwhile for morphological analysis was done to analyse the dispersion of organoclay. The fracture surface of samples was observed by using x-ray diffraction (XRD) to analyze the fracture behaviour of nanocomposite. All datas were recorded by using XRD machine (PANalytical). The diffractometer were scanned in 2θ ranges from 0° until 40°. The peaks were plotted and analyzed.

3. Results and discussion

3.1. Water absorptions

Water absorptions refer to the immersion of nanocomposites samples in the dionized water at room temperature. All the samples were immersed in water for 10 days. Figure 1 shows the percentage of water uptake for all samples at different compositions. The organoclay is hydrophilic properties which is easier to absorb water. This can be observed when the organoclay added at 1% by weight, the percentage of water uptake was increased. Based on the graph, organoclay at 3 wt.% has high percentage of water uptake as compared to the other compositions. As we can observed that when the organoclay loading added above than 5 wt.%, the percentage of water uptake were decreased. This is might be due to the tortuosity pathway of the organoclay in the composite systems which is hindered the water fusion in the nanocomposites. Besides that, the decreased percentage of water uptake might be due to the presence of organoclay which act as nucleating agent could develop the nucleation efficiency and the crystallinity in nanocomposite system then cause the water absorption is less in the composites [6].

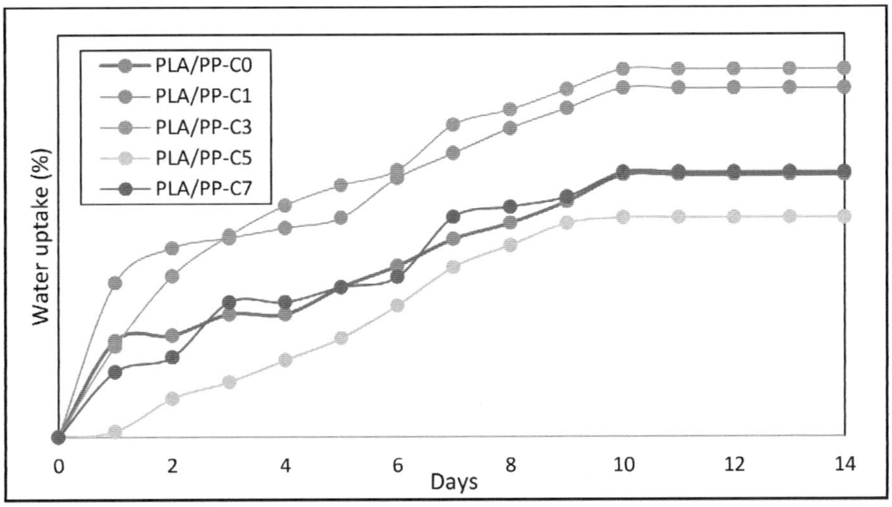

Figure 1. Percentage of water absorptions in nanocomposites of PLA/PP and PLA/PP-C at different loading

3.2. Thermal properties

Differential scanning calorimetry (DSC) were analysed to study the thermal behaviour of nanocomposite. The table 1 shows the tabulated data for glass transition temperature (Tg), melting temperature (Tm) for second heating process which has two peaks Tm_1 and Tm_2 besides crystallization temperature (Tc) during cooling process for this PLA/PP blends with and without organoclay. MAgPP has no effect on the Tm of PLA material. Based on the table 1, Tg_1 values were no significant changes for all compositions even organoclay was added.

Table 1. The thermal properties of PLA/PP and PLA/PP-Clay nanocomposites.

Sample	First heating Tg_1 (°C)	Second heating Tg_2 (°C)	Tm_1 (°C)	Tm_2 (°C)	Tc (°C)
PLA/PP	62.3	54.1	139.9	149.7	113.9
PLA/PP-C1	62.2	51.8	143.6	148.2	111.3
PLA/PP-C3	62.6	51.8	136.7	147.4	111.6
PLA/PP-C5	62.6	51.6	146.6	161.3	111.5
PLA/PP-C7	61.9	49.1	143.1	159.5	112.3

Figure 2 and 3 shows the DSC curve for first and second heating respectively. It was observed that, during the second heating the Tg_2 was shifted to lower temperature from 54.1°C to 51.8°C. The molecular weight loss during the processing due to the polymer chains composition has more free volume compared with neat matrix exhibit the decreasing of Tg in second heating process [7]. Double melting temperature Tm_1 and Tm_2 can clearly seen in both figure 3. It caused by the formation of different structure of crystal which is at high temperature (α-form) and lower temperature (β-form) by the high content of PLA in the composites [8]. The α-form corresponds to most common polymorph and β-form is the imperfect crystal structure. The formation of two peaks at melting temperature also might be due to the corresponds for melting temperature of PLA and also PP material. As shown in

IOP Conf. Series: Materials Science and Engineering **839** (2020) 012017 doi:10.1088/1757-899X/839/1/012017

figure 3, Tm_2 values for nanocomposite were slightly decreased when organoclay was added at 1 wt.% and 3 wt.% because the presence of organoclay could induced the crystallization ability of PLA matrix [9].

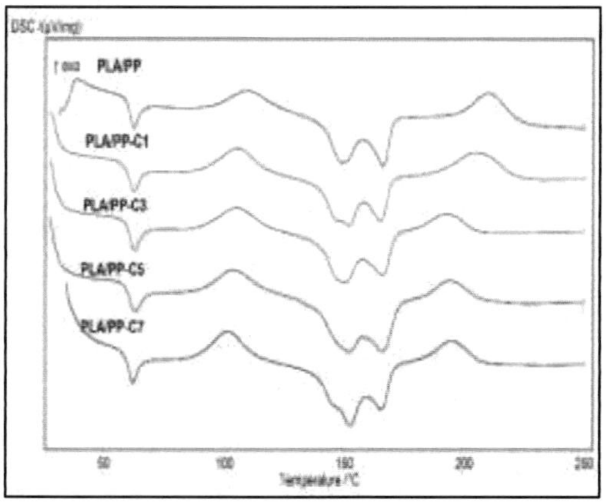

Figure 2. DSC curve first heating for PLA/PP and PLA/PP-C at different loading

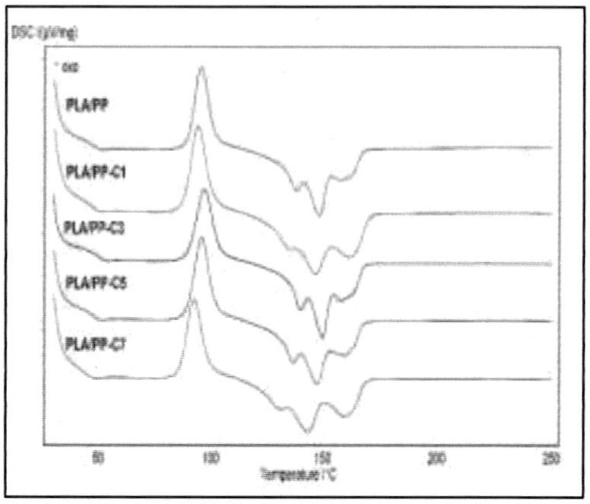

Figure 3. DSC curve second heating for PLA/PP and PLA/PP-C at different loading

DSC cooling curve for PLA/PP and PLA/PP-C nanocomposites are shown in figure 4. The Tc was shifted to lower temperature when organoclay loading was added at 1 wt.% into PLA/PP from 113.9°C to 111.3°C. This might be due to the crystallization happened too slow in PLA matrix to form crystallinity because nanocomposites rapidly cooled from the molten state [10]. The reduction of Tc also can be affected by the compatibilizer which is restricted the formation of crystal of PLA and crystallization process of nanocomposites [11].

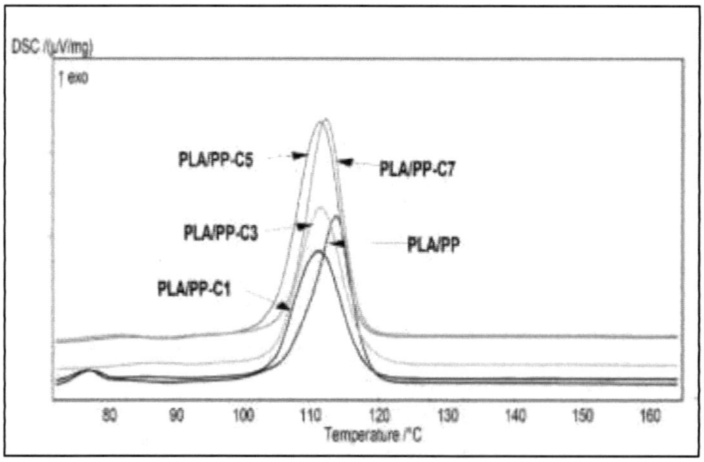

Figure 4. DSC curve for cooling of PLA/PP and PLA/PP-C at different loading

IOP Conf. Series: Materials Science and Engineering **839** (2020) 012017 doi:10.1088/1757-899X/839/1/012017

3.3. Morphological properties

The level of exfoliation and intercalation can be analysed by using XRD analysis. Figure 5 shows the X-ray diffraction pattern of pure organoclay silicate and also the combination of PLA/PP-C with various loading of 1, 3 and 7 wt.%. As shown in the figure 5, actually there were two peaks were found in pure organoclay at 2θ reading under 40°. The diffraction peak which clearly can be seen were at 2θ=7.34° which corresponds to d-spacing 12.05Å and at 2θ=20.26° (4.39Å). In the PLA/PP-C nanocomposites, the diffraction peaks were shifted to the lower degrees of 2θ. The peaks were recorded at 17.68° (5.02Å) for 1 wt.%, 17.45° (5.08Å) for 3 wt.% and 17.5° (5.07Å) for 7 wt.% of organoclay loading as compared to pure organoclay at diffraction peak 2θ=20.26°. This might be due to the factor of intercalation happened with polymer material chains in the clay galleries [12]. The disapperance peak at 2θ below than 10° for the PLA/PP nanocomposites for 1, 3 and 7 wt.% of organoclay might be due to the exfoliated or intercalated structure of organoclay silicates in the specimens.

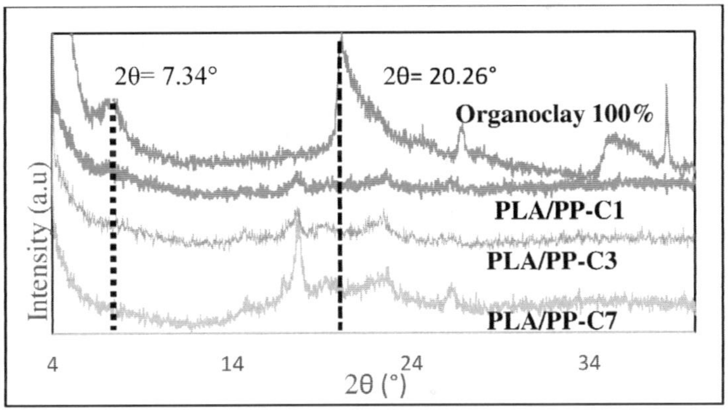

Figure 5. XRD images of PLA/PP-organoclay nanocomposites with (a) 1 wt.% organoclay (b) 3 wt.% organoclay and (c) 7 wt.% organoclay.

4. Conclusions

It was found that, the percentage of water absorptions were increased when organoclay was added at 1 wt.% and when the organoclay loading increased until 7 wt.%, the percentage were decreased. This happened due to the formation of turtuosity pathway of layered silicate which hindered the diffusion of water in the nanocomposites. Meanwhile, in DSC analysis there are formation of two peaks at melting temperature for all compositions. These peaks corresponds to the melting temperature for PLA and PP materials. The crystallization temperature slightly decreased after addition of organoclay. This might be due to the crystallization happened too slow in PLA matrix to form crystallinity because nanocomposites rapidly cooled from the molten state. For XRD analysis, it was found that the reading of 2θ shifted from 20.26° for pure organoclay to the lower degree for organoclay loading at 1, 3 and 7 wt.%. The disappearance peak for all compositions under 10° might be due to the exfoliated and intercalated structure for the organoclay in the PLA/PP blends. In order to develop better behaviour of nanocomposites, it is suggested that the optimum amount of organoclay suitable to be used is within the range 1 until 5% by weight.

5. References

[1] Olalekan S T, Muyibi S A, Shah Q H, Alkhatib M F, Yusof F and Qudsieh I Y 2010 Improving the Polypropylene-Clay Composite Using Carbon Nanotubes as Secondary Filler *Energy Res. J* **1** 68–72

[2] Müller K, Bugnicourt E, Latorre M, Jorda M, Sanz Y E, Lagaron J M, Miesbauer O, Bianchin A, Hankin S, Bolz U, Perez G, Jesdinszki M, Linder M, Scheuerer Z, Castello S and Schmid M 2017 Review on the Processing and Properties of Polymer Nanocomposites and Nanocoatings and Their Applications in the Packaging , Automotive and Solar Energy Fields *Nanomaterials* **74**

[3] Uddin F 2018 Montmorillonite: An Introduction to Properties and Utilization *Web of Science*

[4] Mohsen A H and Ali N A 2018 Mechanical, Color and Barrier, Properties of Biodegradable Nanocomposites Polylactic Acid/Nanoclay *J. Bioremediat. Biodegrad.* **9**

[5] Afroz R, Rahman A, Masud M and Akhtar R 2016 The Knowledge, Awareness, Attitude and Motivational Analysis of Plastic Waste And Household Perspective in Malaysia *Environ. Sci. Pollut. R*

[6] Kord B, Hemmasi A H and Ghasemi I 2011 Properties of PP / wood flour / organomodified montmorillonite nanocomposites *Wood Sci. Technol.* **45** 111–19

[7] Perez J G, Nascimento L, Bou J J, Urquiza E F, Santana O O, Carrasco F and Maspoch M Ll 2010 Influence of Crystallinity on the Fracture Toughness of Poly (lactic acid)/ Montmorillonite Nanocomposites Prepared by Twin-Screw Extrusion *J. Appl. Polym.* **120** 896-905

[8] Fukushima K., Tabuani D and Camino G 2009 Nanocomposites of PLA and PCL based on montmorillonite and sepiolite *Mater. Sci. Eng. C* **29** 1433–41

[9] Luo Y, Liu H Y, Zhang G Z and Qu J P 2017 Morphological and thermal properties of PLA / OMMT nanocomposites prepared via vane extruder *Mater. Sci. Eng.* **213**

[10] Fukushima K, Tabuani D, Arena M, Gennari M and Camino G 2013 Effect of clay type and loading on thermal , mechanical properties and biodegradation of poly (lactic acid) nanocomposites *React. Funct. Polym.* **73** 540–49

[11] Jamshidian M, Tehrany E A, Imran M, Jacquot M and Desobry S 2010 Poly-Lactic Acid : Production , Applications , Nanocomposites, and Release Studies *Comp. Rev. Food Sci Food Saf.* **9** 552–71

[12] Salehiyan R, Ray S S, Bandyopadhyay J and Ojijo V 2017 The Distribution of Nanoclay Particles at the Interface and Their Influence on the Microstructure Development and Rheological Properties of Reactively Processed Biodegradable Polylactide/Poly(butylene succinate) Blend Nanocomposites *Polymers* **350**

Acknowledgements

The authors would like to thank Universiti Teknologi MARA, Malaysia and the Institute Research Management & Innovative (IRMI) for the support for this research. This project was funding under the Fundamental Research grant Scheme (FRGS), 600-IRMI/FRGS 5/3 (070/2019).

IOP Conf. Series: Materials Science and Engineering **839** (2020) 012018 doi:10.1088/1757-899X/839/1/012018

Experimental study on the shear behavior of GFRP reinforced concrete beams strengthened using CFRP sheets

M Talha Junaid, Abdul Saboor Karzad, Abdalla Elbana

Department of Civil and Environmental Engineering, University of Sharjah, Sharjah, UAE

Abstract. This paper presents the results of an experimental investigation on the shear capacity of reinforced concrete beams. The beams under consideration are reinforced with glass fiber-reinforced polymer (GFRP) bars. Also, the effect of shear strengthening on such beams is studied. For this purpose a total of four beams were tested where two served as control beams, while the other two were strengthened using U shape carbon fiber-reinforced polymer (CFRP) sheets, bonded externally. Each beam is 2.4 m long, simply supported and subjected to a four-point load under displacement control mode. The strengthened beams showed a noticeable increase in the load-bearing capacity of the specimen. The failure load recorded of these specimens was 208 kN, an increase of 25%. Moreover, the mode of failure changed from pure shear to concrete crushing in the compression zone. The results indicate that using CFRP sheets to strengthen GFRP reinforced beams in shear is a viable option. However, the results also suggest that the known adverse interaction between EB-CFRP and conventional steel stirrups could be applicable between EB-CFRP and GFRP shear reinforcement in strengthened beams. Therefore the predicted EB-CFRP shear capacity by ACI 440.2R turned to be unconservative when compared to the experimental results.

1. Introduction

Concrete and steel as being commonly used construction materials have an extensive track record for over 150 years. Design methods and applications of these materials have been already established and advanced. However, the fiber-reinforced polymers (FRP) is a newly introduced material to the construction industry. And some of the behavioral aspects of this material are not yet established. FRP is a composite material made of polymer matrices reinforced with fibers. The fibers are usually glass, carbon, aramid, or basalt. While polymer can be epoxy, vinyl ester, or polyester thermosetting plastic [1]. Such materials (FRP) can be used as internal reinforcement in form of reinforcement bars or as an external reinforcement in form of warp fabric sheet or strip for strengthening or repair of existing deficient structural members [2]–[4]. FRPs are known for their durability, non-corrosiveness, lightweight, high strength to weight ratio, nonmagnetic, and ease in usage [5].

Under severe environmental conditions structures are prone to deterioration particularly due to corrosion of traditional steel reinforcement. To tackle this drawback associated with steel reinforcement, the FRP bars have emerged as a possible alternative to steel bars. The use of GFRP rebars as reinforcement has been intensively studied in the literature [6]–[9]. Some of these studies undertook the bonding behavior between the FRP reinforcement and concrete. While other researchers studied the shear and the flexural capacity of structural members containing these types of reinforcements [10]–[12]. For instance, Kassem et al. [7] evaluated the flexural behavior of 24 beams specimens reinforced with different types of FRP bars. It was concluded that the sand coated bars had better bond compared to the bars with ribbed surface. All the tested beams eventually failed due to concrete crushing which is the preferred mode of failure in FRP reinforced beams. In addition the paper reported that the CSA S806-02 model was better fitting the experimental data.

Owing to its superior durability, the FRP strip can be externally bonded to a concrete member to restore or even improve its load-carrying capacity. For instance, it can be used for upgrading the bridge capacity in case of higher traffic volume, seismic retrofitting, or even adding another floor to an existing building [13]. Although the use of FRP as an internal and external reinforcement is widely reported in a conventional reinforced member [3]. The use of glass fiber reinforced polymer (GFRP) stirrups as shear reinforcement of a beam is not well studied. On the other hand, strengthening the beams using the growing technique of externally bonded carbon fiber reinforced polymer (EB-CFRP) sheets has been widely studied in the literature. However, shear strengthening using EB-CFRP is still a matter of research [14-17]. Moreover, there does not exist a study in the literature that investigates the shear strengthening of GFRP reinforced concrete beams.

In the current study beams with GFRP bars used as longitudinal and shear reinforcement is studied. One of the two objectives was to investigate the use of GFRP bars as replacement to the conventional steel which is prone to corrosion. The studied beams were intentionally designed to be shear deficient for achieving the second objective of the study which is shear strengthening using EB-CFRP technology. A total of four beams which are part of an ongoing extensive research project, were considered in this experimental research paper.

2. Experimental Program

2.1. Materials and Test Specimens

The concrete mix used to cast the test specimens was OPC based with an average 28 days compressive strength of 50 MPa. The mix proportions of the concrete used to cast the beams are shown in Table 1. The test specimens were reinforced with GFRP longitudinal bars and stirrups. The GFRP longitudinal bars and stirrups were manufactured and provided by Pultron composites company under the product name MateenBar (f^{*}_{fu} = 690 MPa, E_f= 51 ±2.5 GPa, and ε^{*}_{fu}= 0.0135,). For shear strengthening of the test specimens, a uni-directional carbon fiber reinforced polymer (CFRP) fabric sheets provided by Conmix LTD was used (f_u = 4000 MPa, E_f = 230 GPa, and ε_u = 1.8 %).

The test specimens were 2400 mm long beams with a rectangular cross-section of 200 x 300 mm, and an effective depth of 250 mm and an active span of 2100 mm (see Figure 1). The control beams were designed as per the ACI 440.1R-06 guideline such that the shear failure will be the dominant mode of failure even in addition to strengthening. Therefore, the beams were over reinforced in flexural by placing 3#22 GFRP bars in tension zone and also 2#10 in the compression zone. While only minimum shear reinforcement was considered by using #8 GFRP stirrups spaced at 200mm c/c. A total of four beams were cast where two were tested to failure without strengthening served as control beams and two were strengthened in shear prior to testing.

Table 1. Concrete mix proportions

Contstituent	Quantity	Unit
OPC	365	kg/m^3
20mm aggregates	646	kg/m^3
10mm aggregates	290	kg/m^3
5mm fine aggregates	530	kg/m^3
Dune sand	369	kg/m^3
Superplasticizer	4	Liter/m^3
Water	175	Liter/m^3

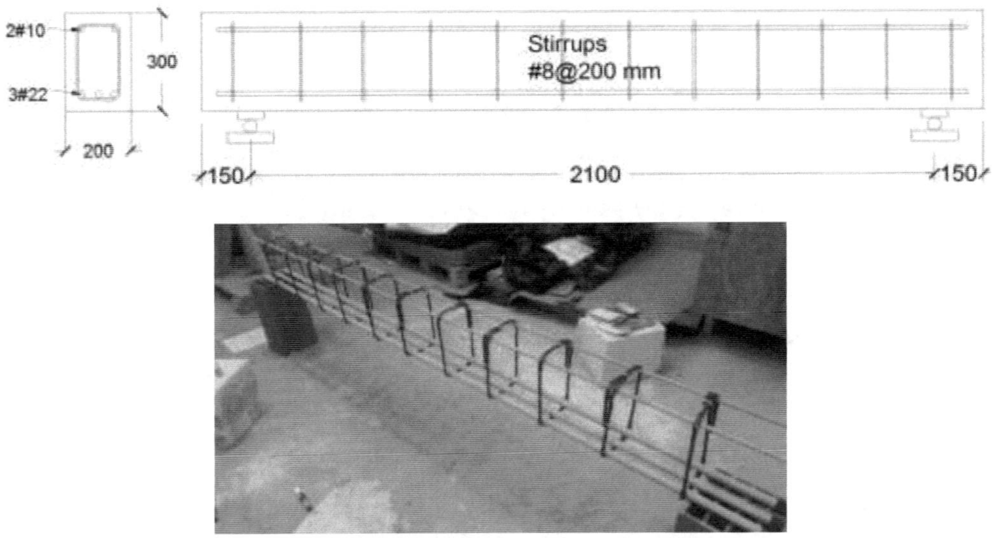

Figure 1. Test specimens' dimensions and reinforcement details

2.2. Strengthening Methodology

In this study, two out of four GFRP reinforced beams were strengthened in shear using one layer of EB-CFRP discrete sheets (strips) in a U shape configuration. The strips were 130 mm wide and spaced at 200 mm center to center of the strips (net spacing of 70 mm). The stirps were bonded vertically to the sides and soffit of the beams by adopting the wet layup application. It should be mentioned that prior to bonding and encapsulation of the CFRP strips, surface preparation was executed. Therefore, the surface of the beams was first grinded, and surface irregularities were leveled using mortar then a primer coat was applied. In addition the corner of the beams were chamfered to prevent stress concentration. Figure 2 shows the layout of strengthening along with a photo showing the application of EB-CFRP on to the beams.

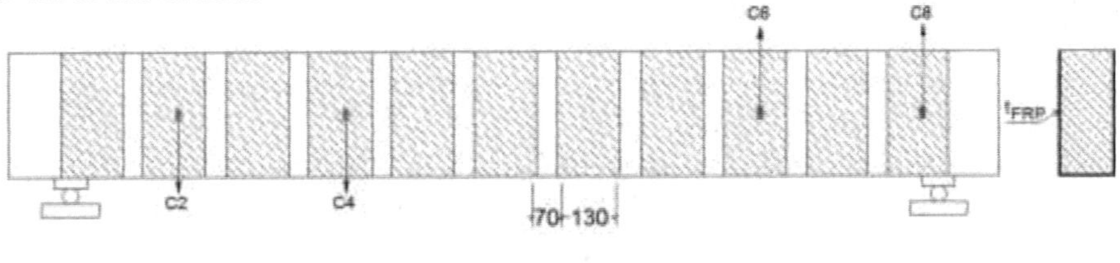

(a) Schematic layout of EB-CFRP strengthening

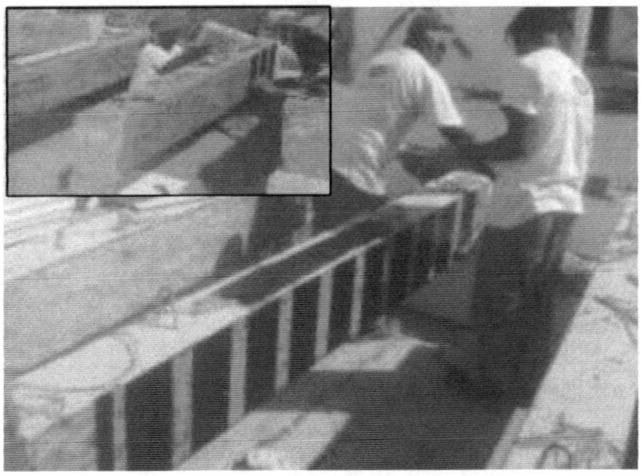

(b) Application of EB-CFRP sheets to the studied specimens

Figure 2. Wrapping configuration and process

2.3. Test Setup and Instrumentation

The beams were tested under a four-point static loading in a simply supported configuration as shown in Figure 3. The load was applied using a hydraulic actuator with a rate of 0.01 mm/sec in a displacement control mode. A linear variable displacement transducer (LVDT) was deployed under the beam in the midspan to measure the beam deflection during test. Meanwhile, electrical resistance surface mounted strain gauges were attached to EB-CFRP sheets to measure the strain development under the load.

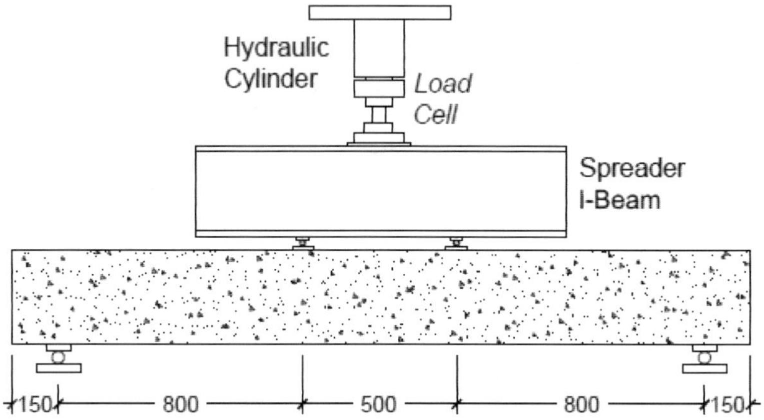

Figure 3. Schematic diagram of the test set up

2.4. Test Results

Figure 4 shows a photo of the control beam and a strengthened beam after failure. As can be seen in Figure 4a, the control beams failed in shear evident from diagonal shear crack on both the shear spans as expected. The control beams resisted an average maximum load of 165 kN and a maximum deflection of 25 mm (see Figure 5) prior to collapse due to the major shear crack. However, the strengthened beams failed due to concrete crushing at the compression zone. The average maximum load and maximum deflection for the strengthened beams were 208 kN and 45 mm respectively as shown in Figure 5.

(a) Photo of control (Ctr) beam after failure

(b) Photo of strengthened (Str) beam after failure

Figure 4. Photos of tested specimens after failure

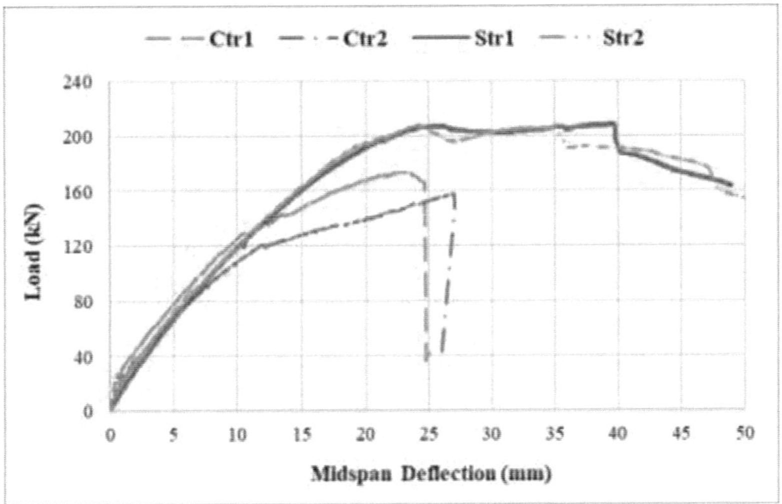

Figure 5. Load-deflection diagram for all the tested samples

3. Discussion

3.1. Shear Strength Increment and Failure Modes

The shear strength of the beams which is half of the registered applied ultimate load are presented in Table 2 along with maximum deflection and failure modes. The shear strength of the reference beams is recorded directly from the testing device. Similarly, for the total shear capacity of strengthened beams. However, the contribution of the EB-CFRP sheets to the shear strength can be calculated by subtracting the shear strength of the control beams from that of the strengthened beams.

Table 2. Summary of test results

ID	Ultimate Load P (kN)	Shear Capacity Vn (kN)	Increase in Shear Capacity (%)	Deflection at Failure (mm)	Failure mode
Ctr1	173	87	NA	24	Diagonal Shear Crack
Ctr2	157	79	NA	27	Diagonal Shear Crack
Str1	208	104	20	45	Concrete Crushing
Str2	208	104	32	45	Concrete Crushing

3.2. Guidelines Predictions

The internal shear reinforcement used were based on the prediction model of ACI 440.1R:

$$V_n = V_c + V_f \tag{1}$$

Where, V_c is the shear contribution of the concrete for an FRP-reinforced member:

$$V_c = \frac{2}{5}\sqrt{f_c'}\,b_w c \tag{2}$$

$$c = kd \tag{3}$$

$$k = \sqrt{2\rho_f n_f + (\rho_f n_f)^2} - \rho_f n_f \tag{4}$$

$$\rho_f = {A_f}/{b_w d} \tag{5}$$

And, V_f is the FRP stirrups contribution in shear:

$$V_f = \frac{A_{fv} f_{fv} d}{S} \tag{6}$$

A minimum shear reinforcement was provided:

$$A_{fv,min} = 0.35\frac{b_w s}{f_{fv}} \tag{7}$$

$$S = \frac{\emptyset A_{fv} f_{fv} d}{V_u - \emptyset V_c} \tag{8}$$

$$f_{fv} = 0.004E_f \leq f_{fb} \tag{9}$$

Where b_w is the web width, c is cracked transformed section neutral axis depth, d is the distance from the compression zone to tension reinforcement, S is spacing of FRP stirrups, f_{fv} is the tensile strength of FRP, E_f is the elastic modulus of FRP reinforcement, ρ_f is FRP reinforcement ratio, n_f is modular ratio, and f_{fb} the allowable stress in the stirrup at the location of a bend.

It was decided to strengthen the beams such that the shear strength of the beams will be increased by at least 30%. To obtain the targeted increase in shear strength, ACI 440.2R guideline was adopted to design the strengthening configuration and specify the required width and spacing of CFRP sheets. Only one layer of EB-CFRP sheets with a thickness of 0.168 mm, width of 130 mm, and spacing of 200 mm was used. The predicted contribution of EB-CFRP sheets to shear strength of the beams is calculated using following equations:

$$V_{frp} = \psi \cdot \frac{A_{frp} \cdot f_{frp,e} \cdot (\cos\alpha + \sin\alpha) \cdot d_{frp}}{s_{frp}} \tag{10}$$

$$A_{frp} = 2 \cdot n \cdot t_{frp} \cdot w_{frp} \qquad f_{frp,e} = \varepsilon_{frp,e} \cdot E_{frp} \tag{11}$$

$$\varepsilon_{frp,e} = \begin{cases} 0.004 \le 0.75 \cdot \varepsilon_{frp,u};\ \textit{for full wrapping} \\ k_v \cdot \varepsilon_{frp,u} \le 0.004;\ \textit{for side or U jacketing} \end{cases} \quad (12)$$

$$k_v = \frac{k_1 \cdot k_2 \cdot L_e}{11900 \cdot \varepsilon_{frp,u}} \le 0.75 \qquad L_e = \frac{23300}{(n \cdot t_{frp} \cdot E_{frp,u})^{0.58}} \quad (13)$$

$$k_1 = \left(\frac{f_c'}{27}\right)^{2/3} \qquad k_2 = \begin{cases} \dfrac{d_{frp} - 2L_e}{d_{frp}};\ \textit{for side bonding} \\ \dfrac{d_{frp} - L_e}{d_{frp}};\ \textit{for U jacketing} \end{cases} \quad (14)$$

Where, Ψ is the reduction factor equal to 0.85, A_{frp} is the area of CFRP sheet, $f_{frp,e}$ is CFRP effective stress at failure, d_{frp} is the CFRP effective height, S_{frp} is the spacing between two adjacent FRP discrete sheets, α is the angle of fiber orientation, n is the number of CFRP layers applied, t_{frp} is the thickness of CFRP sheet, w_{frp} is the width of one discrete sheet, $\varepsilon_{frp,e}$ is CFRP effective strain at failure, E_{frp} is FRP elastic modulus, $\varepsilon_{frp,u}$ is CFRP ultimate strain, L_e is the CFRP effective length, k_v is bond dependent coefficient, k_1 is modification factor to account for concrete strength, and k_2 is modification factor to account for FRP wrapping scheme.

Table 3 shows the experimental values of shear strength contribution by concrete, GFRP, and EB-CFRP along with the shear capacity predictions by ACI models namely ACI 440.1R and ACI 440.2R. Comparsion of the experimental values with that of the model predictions reveals that the ACI 440.1R reasonably predicts the shear strength of control beams with a ratio (V_{exp}/V_{pre}) of 1.2. However, the ACI 440.2R model overestimated the shear contribution provided by EB-CFRP sheets. The experimental value of EB-CFRP shear contribution was calculated by simple subtraction of V_{nc} (V_C+V_f) from V_{ns} ($V_c+V_f+V_{FRP}$). This turned to be an average of 21 kN while the model prediction was 42 kN. Therefore, the ratio of V_{exp}/V_{FRP} is less than 1 for the case of strengthened beams. Which indicates that there is possibility of an interaction between GFRP stirrups and CFRP sheets similar to the case of normal steel reinforced concrete strengthened beams that affects the model predictions. However, further investigation is needed to address this issue.

Table 3. Comparison of guidelines predictions with experimental results

ID	Shear Capacity V_{exp} (kN)	ACI 440.1R Model Predictions $V_{nc,pre}$ $(V_c+V_f)_{pre}$ (kN)	ACI 440.2R Model Predictions $V_{FRP,pre}$ (kN)	ACI Model Predictions $V_{ns}(V_c+V_f+V_{FRP})$ (kN)	V_{exp} / $V_{ns,pre}$
Ctr1	87	72	-	72	1.2
Ctr2	79	72	-	72	1.1
Str1	104	72	42	114	0.9
Str2	104	72	42	114	0.9

4. Conclusions

This work looks at the shear behavior of GFRP reinforced concrete beams and the applicability of using CFRP sheets for strengthening such composite syustems. The following conclusions can be drawn from this work:

1. GFRP shear reinforcement is a viable alternate to traditional steel reinforcement in shear.

2. The prediction model as put forward by ACI 440.1R for GFRP reinforced concrete beams gives accurate prediction to the ultimate shear load capacity.

3. Discrete CFRP sheets may be used to strengthen GFRP-concrete composite systems in shear. The current work indicate that an average strength gain of 25% was achieved using a single layer of CFRP sheet. The deflection behavior of the strengthened beams was also altered significantly, with both specimens behaving in a ductile manner with a maximum mid-span deflection of 40mm (an increase of 60%).

4. The strengthening model as proposed by ACI 440.2R over predicts the shear capacity provided by EB CFRP sheets. This may indicate a possible interaction between the GRRP shear stirrups and the CFRP sheets, as also reported for traditional steel shear reinforced beams. However further investigation is need to investigate this interaction.

5. References

[1] ACI, 440R State-of-the-Art Report on Fiber Reinforced Plastic (FRP) Reinforcement for Concrete Structures (R.2002) ACI 440R-96

[2] ACI, 440.2R Guide for the Design and Construction of Externally Bonded FRP Systems 2008 ACI 440.2R

[3] Pendhari S S, Kant T, and Desai Y M 2008 Application of polymer composites in civil construction: A general review *Compos. Struct* **84** pp 114–24

[4] Alkhrdaji T et al. 2006 ACI Committee 440. Guide for the design and construction of concrete reinforced with FRP bars ACI 440.1R-06.

[5] Triantafillou T C 1999 Shear Strengthening of Reinforced Concrete Beams Using Epoxy-Bonded FRP Composites *ACI Struc J.* **95** 107–15

[6] Ben Ouezdou M, Nationales J and Beton D U 2017 On the internal and external FRP reinforcement of RC beams *International Congress on Materials & Structural Stability*

[7] Kassem C, Farghaly A S and Benmokrane B, 2011 Evaluation of flexural behavior and serviceability performance of concrete beams reinforced with FRP bars *J. Compos. Constr.* **15** 682–95

[8] Tavares D H, Giongo J S and Paultre P 2008 Behavior of reinforced concrete beams reinforced with GFRP bars *Rev. IBRACON Estruturas e Mater.* **1** 285–95

[9] Alsayed S H, Al-Salloum Y Aand Almusallam T H 2000 Performance of glass fiber reinforced plastic bars as a reinforcing material for concrete structures *Compos. Part B Eng.* **31** 555–67

[10] Achillides Z and Pilakoutas 2004 Bond Behavior of Fiber Reinforced Polymer Bars under Direct Pullout Conditions *J. Compos. Constr.* **8** 173–81

[11] AL-mahmoud F, Castel A, François R and Tourneur C 2007 Effect of surface pre-conditioning on bond of carbon fibre reinforced polymer rods to concrete,*Cem. Concr. Compos* **29** 677–89

[12] Soltanzadeh F, Mazaheripour H and Barros J 2013 Shear Capacity of HPFRC Beams Flexurally Reinforced with Steel and Prestressed GFRP Bars *Frprcs-11* pp. 1–10

[13] Bousselham Aand Chaallal O 2006 Effect of transverse steel and shear span on the performance of RC beams strengthened in shear with CFRP *Compos. Part B Eng.* **37** 37–46

[14] Kim Y, Ghannoum W M and Jirsa J 2015 Shear behavior of full-scale reinforced concrete T-beams strengthened with CFRP strips and anchors *Constr. Build. Mater.* **94** 1–9

[15] Li W, Hu C, Pan Z, Peng W, Yang Yand Xing F 2018 A proposed strengthening model

considering interaction of concrete-stirrup-FRP system for RC beams shear-strengthened with EB-FRP sheets *J. Reinf. Plast. Compos.* **37** 685–700

[16] Karzad A S, Leblouba M, Al Toubat S and Maalej M. 2019 Repair and strengthening of shear-deficient reinforced concrete beams using Carbon Fiber Reinforced Polymer *Compos. Struct.* **223**

[17] Altoubat S, Karzad A S, and Maalej M 2018 Strengthening of Damaged Reinforced Concrete Beams Using Externally Bonded Fiber Reinforced Polymer *The 13th International Symposium on Fiber-Reinforced Polymer Reinforcement for Concrete Structures.*

Acknowledgments

The authors would like to acknowledge the support provided by Pultron Composites company for providing the GFRP bars and Conmix LTD for supplying the CFRP sheets. The project was funded by grant number (1602040128-P) from University of Sharjah. The help of Sustainable Construction Material & Structural System research group and contribution of Eng. Ahmed Shweiki during testing is also appreciated.

Study of the hardness and microstructure profiles of SKD61 steel plunger tip (local material) after plasma nitriding

Usman Sudjadi and R V Sinaga
Center for Nuclear Fuel Technology – National Nuclear Energy Agency
E-mail: umardjasipan@gmail.com

Abstract. The Hardness and Microstructure Profiles of SKD61 Steel Plunger Tip (local material) after Plasma Nitriding have studied. To improve SKD61 quality for high temperature use, a plunger tip can be hardened through surface treatment using a plasma/ion nitriding technique. To test the success of the surface treatment, the nitriding temperature was varied from 350ºC to 550ºC with step for 50 ºC for the nitriding process completed for 10 hours. The results, it can be seen that the optimum surface hardness was achieved when the nitriding temperature was 500ºC with Vickers hardness (HV) was 955.29 HV with a depth of up to 100 μm from the surface of the specimen. At such a nitriding temperature, the hardness increased 1.79 times in comparison with that after hardening only, and it increased 2.076 times in comparison with that of the initial hardness. From the results of the study, it can be concluded that plasma nitriding temperature affects the amount of nitrogen concentration diffused into the substrate material. Nitride layer formed was possibly of Fe N, Cr_2N, VN, δ Fe, and γ Fe.

Keywords: Plasma/Ion Nitriding, SKD61 Steel, High Pressure Die Casting.

1. Introduction

A plunger tip is an important component in a high pressure die casting of cold chamber type that serves as an aluminum fluid booster in the cold sleeve to fill in the mold cavity. The plunger tip is always rubbing against the inner wall of the cold sleeve or in direct contact with the aluminum metal fluid poured in the cold sleeve or chamber. To keep the expansion steady, there is a cooling water circulation inside the plunger tip. The water circulation goes through a small pipe connecting the plunger tip and the hydraulic cylinder.

Figure 1. A 65 x 90 mm Plunger Tip

With a high pressure, aluminum fluid is pushed by the plunger tip in to the mold cavity/dies. Friction between the surfaces results in wear and tear. Besides the wear problem due to friction between metallic surfaces, the plunger tip also warms up due to direct contact with aluminum fluid during operation. For a durable plunger tip, it needs a material that is strong, wear-resistant, and it does not experience a decrease in hardness at high temperatures. Typically, a plunger tip in a high pressure die casting machine is made of hardened SKD61 steel with a hardness of 457.6 - 499.2 HV For high temperature uses, hardening treatment alone is not enough. An additional surface treatment needs to be done on the plunger tip material to improve its quality at high temperature operation. A research had been conducted to observe the effect of plasma nitriding on the hardness and the microstructure profiles of SKD61 steel as a plunger tip component of high pressure die casting machines.

2. Research methodology

The specimens used in this research were SKD61 steel, each had a size of 22 x 20 mm. The steel was the material for a 65 x 95 mm plunger tip obtained from PT ASSAB Steels Indonesia. SKD61 steel is equivalent to ASSAB 8407 2M (a standard material of PT ASSAB Steels Indonesia) which has the following composition: C 0.39 - Si 1.0 - Mn 0.4 - Cr 5.3 - V 0.9 - Mo 1.3. The ASSAB 8407 2M material has an initial hardness of 460 HV. At such hardness, the process of forming or cutting the metal is easier. Generally after the cutting or machining is done, the material undergoes hardening process to achieve the standard level appropriate for a plunger tip. The typical hardness of SKD61 material is 499.2-572 HV. In this research, each specimen was cleaned with distilled water. After that, each one was cleaned with acetone. The natural oxide layer on the surface of the specimen was etched with HF for one to three minutes. Each specimen was cleaned again with distilled water. Then all specimens were sprayed with nitrogen gas. After that, the specimens were hardened on their surface with DC plasma nitriding [see Fig. 2] at temperatures of 350°C, 400°C, 450°C, 500°C, and 550°C for 10 hours. Nitrogen gas used in the experiment was of high purity. Then specimens were cut, mounted, polished, and etched for microstructure observation with an optical microscope. The surface hardness of the specimens were tested with Vickers AKASHI MVK micro-machine - E load (g) 10 - 1000. The load was 100 gram-force.

Figure 2. The DC plasma nitriding (or nitrocarburizing) apparatus for this experiment.

This tool can be used for both nitriding and nitrocarburizing processes. When it is used for nitriding process, the gas source that produces carbon atoms is turned off, and only the nitrogen gas source is opened.

3. Results and Discussion

3.1. Initial SKD61 Composition and Hardness Test

Based on the supplier's material certificate issued by PT ASSAB Steels Indonesia No: 076/8407 2M/13 on 14 January 2013, the composition of SKD61 or ASSAB 8407 2M was as follows:
C 0.39 - Si 1.0 - Mn 0.4 - Cr 5.3 - V 0.9 - Mo 1.3. The material hardness was 460 HV.

3.2. Hardness Testing of SKD61 Specimens

The hardness test was carried out using Vickers AKASHI MVK micro-machine - E load (g) 10 - 1000. The load was 100 gram-force. Vickers testing was carried out on each specimen after hardening and plasma nitriding at 350°C, 400°C, 450°C, 500°C, and 550°C.

3.2.1. SKD61 Hardness after Hardening Process.

IOP Conf. Series: Materials Science and Engineering **839** (2020) 012019 doi:10.1088/1757-899X/839/1/012019

Hardness testing with Vickers technique for SKD61 specimens after hardening was carried out by tracking at depths of ranging from 0.02 mm to 2 mm from the specimen's surface and using 100 gram-force load as shown in Figure 3.

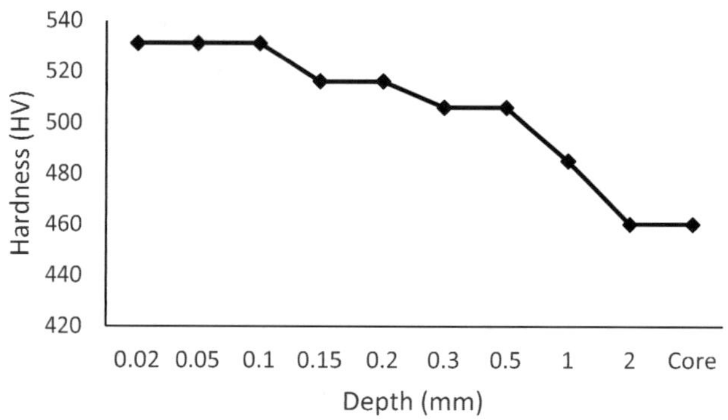

Figure 3. Hardness of the specimen initial state before surface hardening with DC plasma nitriding

For the hardness test, ten observation points were determined, i.e. at the depth of: (1) 0.02 mm, (2) 0.05 mm, (3) 0.10 mm, (4) 0.15 mm, (5) 0.20 mm, (6) 0.30 mm, (7) 0.50 mm, (8) 1.00 mm, (9) 2.00 mm, and (10) beyond 2.00 mm (core). Measurement results showed that the hardness of SKD61 steel after the plasma nitriding increased from the initial hardness of 460 HV to 531.32 HV. Such a difference in hardness was due to the hardening process carried out in vacuum where the cooling was done with N_2 medium. At the surface that is the first observation point, the hardness was 531.32 HV, while at the depth of 2.00 mm (core), the hardness was 460.27 HV.

3.2.2. SKD61 Hardness after Plasma Nitriding at 350°C.

Hardness testing with the Vickers technique for SKD61 specimens after plasma nitriding was carried out by tracking at the depth from 0.02 mm to the 0.3 mm from the surface and using 100 gram-force load as shown in Figure 4.

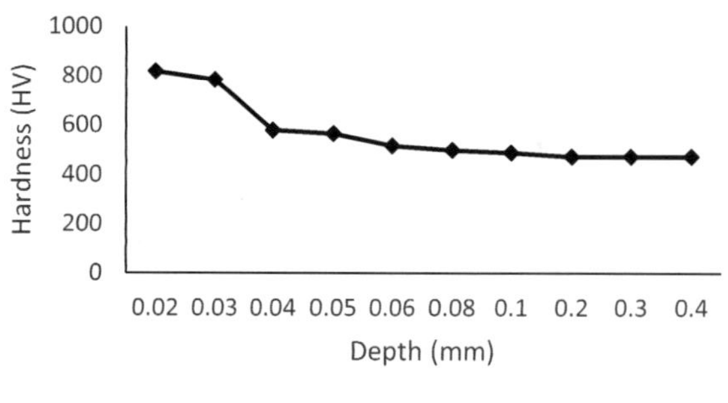

Figure 4. Hardness of the specimen after 10-hour plasma nitriding at 350°C.

For the hardness test, ten observation points were determined, i.e. at the depth of: (1) 0.02 mm, (2) 0.03 mm, (3) 0.04 mm, (4) 0.05 mm, (5) 0.06 mm, (6) 0.08 mm, (7) 0.10 mm, (8) 0.20 mm, (9) 0.30 mm,

IOP Conf. Series: Materials Science and Engineering **839** (2020) 012019 doi:10.1088/1757-899X/839/1/012019

and (10) beyond 0.30 mm (core section). Measurement results showed that the hardness of SKD61 steel after plasma nitriding increased from its hardness after hardening of 531.32 HV to 818.20 HV.

3.2.3. SKD61 Hardness after Plasma Nitriding at 400°C.

Hardness testing with the Vickers technique for SKD61 specimens after plasma nitriding was carried out by tracking at the depth from 0.02 mm to the 0.4 mm from the surface and using 100 gram-force load as shown in Figure 5.

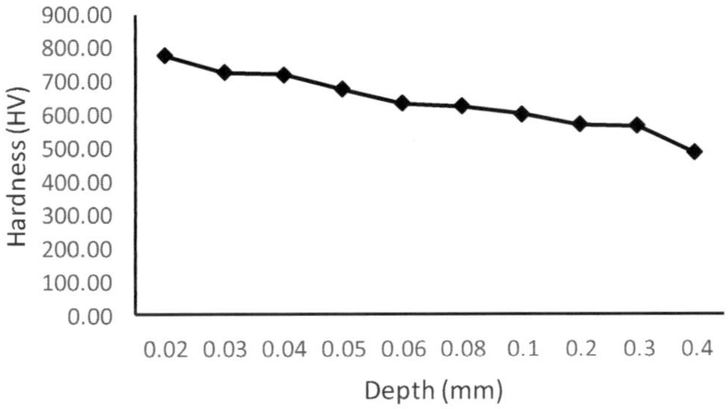

Figure 5. Hardness of the specimen after 10-hour plasma nitriding at 400°C

The hardness at the surface (at 0.02 mm depth) of the specimen treated with plasma nitriding at 400°C was lower than that at 350°C, that was from 818.20 HV to 776.70 HV. However, the hardness at 400°C nitriding temperature was more evenly distributed than that at 350°C nitriding. This is probably due to the accumulation of nitrogen atom layers on the surface of the specimen resulting in decreased hardness. The effect of the buildup of metal nitride layers on the surface can lead to a decrease in hardness. This phenomenon can occur if there is no equilibrium between the diffusion rate of N atoms on the surface and the deposition rate. The diffusion rate of N atoms on a metal surface is strongly influenced by the temperature of the substrate while the nitrogen ion deposition rate is affected by nitriding pressure. If the deposition rate is too large but diffusion and reaction of nitrogen atom (N) with substrate (Fe, Cr, Al, Mo, W, and Ti) cannot follow, it is possible to accumulate nitrogen atoms on the surface causing surface hardness to rise but will go down [1]. The results showed that the hardness decreased along with the depth until it reached the initial hardness of the SKD61 specimen. The optimum hardness was reached at a depth of 0.04 mm or 4μm. This indicates that the deeper the concentration of nitrogen at a lower concentration and the possibility of the occurrence of metal nitride compounds, the fewer the resulting violence decreases. On the surface, the hardness was 776.70 HV, while at a depth of 0.40 mm the hardness was 485.40 HV.

3.2.4. SKD61 Hardness after Plasma Nitriding at 450°C.

Hardness testing with the Vickers technique for the SKD61 specimens after plasma nitriding was carried out by tracking the depth from 0.02 mm to the tracking depth of 0.4 mm from the surface and 100 gram-force load as shown in Figure 6.

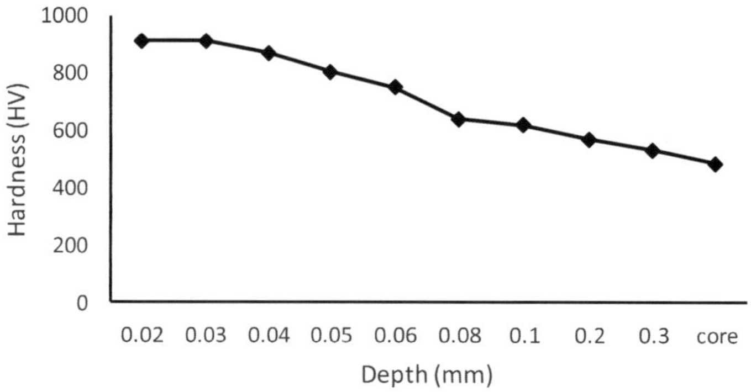

Figure 6. Hardness of the specimen after 10-hour plasma nitriding at 450°C

For the hardness test, ten observation points were determined, i.e. at the depth of: (1) 0.02 mm, (2) 0.03 mm, (3) 0.04 mm, (4) 0.05 mm, (5) 0.06 mm, (6) 0.08 mm, (7) 0.10 mm, (8) 0.20 mm, (9) 0.30 mm, and (10) beyond 0.40 mm (core section). Measurement results showed that at the surface, the hardness of SKD61 steel after 450°C plasma nitriding increased from its hardness after 400°C plasma nitriding, that is from 776.70 HV to 911.58 HV. The results showed that the hardness decreased along with the depth until it reached the initial hardness of the SKD61 specimen. The optimum hardness was reached at a depth of 0.03 mm or 3 μm. This indicates that the deeper the concentration of nitrogen at a lower concentration and the possibility of the occurrence of metal nitride compounds, the fewer the resulting violence decreases. On the surface, the hardness was 911.58 HV, while at a depth of 0.40 mm the hardness was 485.42 HV. This indicates that the change in hardness at 450°C nitriding includes up to 0.40 mm depth until the hardness is the same as the initial hardness of the material before hardening.

3.2.5. SKD61 Hardness after Plasma Nitriding at 500°C.
Hardness testing with the Vickers technique for the SKD61 test sample after the plasma nitriding process was carried out by tracking the depth from 0.02 mm to the tracking depth of 0.4 mm from the sample surface and 100 gram-force load as shown in Figure 7. For the hardness test, ten observation points were determined, i.e. at the depth of: (1) 0.02 mm, (2) 0.03 mm, (3) 0.04 mm, (4) 0.05 mm, (5) 0.06 mm, (6) 0.08 mm, (7) 0.10 mm, (8) 0.20 mm, (9) 0.30 mm, and (10) beyond 1.0 mm (core section).

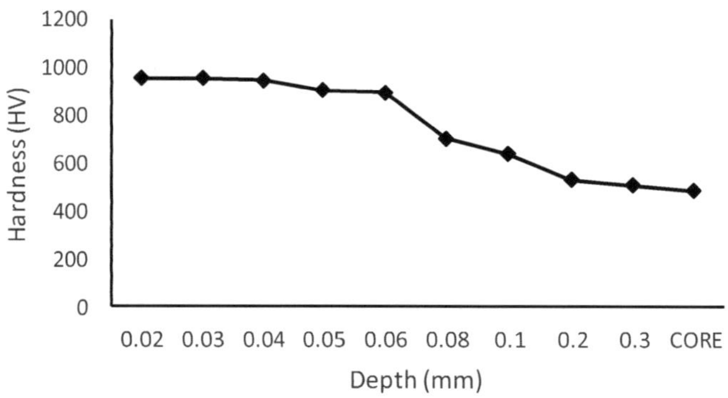

Figure 7. Hardness of the specimen after 10-hour plasma nitriding at 500°C.

Based on the results of the measurement of the hardness shown in the graph above, the measurement results of SKD61 material after the plasma nitriding process increased from the previous hardness at nitriding temperature of 450ºC which was 911.58 HV to 955.29 HV. Based on the test results show that the deeper the surface of the hardness value decreases until it reaches the same hardness as the initial material of the SKD61 specimen. The optimum hardness value is reached at 0.05 mm or 5µm depth. This indicates that the deeper the concentration of nitrogen at a lower concentration and the possibility of the occurrence of metal nitride compounds, the fewer the resulting violence decreases. On the surface edge of the first tracking sample it obtained hardness of 955.29 kgf/mm^2 , while at a depth of 1.00 mm (core) the hardness value obtained was 485.42 kgf/mm^2. This indicates that the resulting hardness change in the nitriding process temperature of 500ºC covers a depth of 1mm until the resulting hardness is the same as the initial hardness of the material before hardening, which is 460 HV at a depth of 2 mm from the metal surface. Based on the results of the measurement of the hardness shown in the graph above, the measurement results of SKD61 material after the plasma nitriding process increased from the previous hardness at nitriding temperature of 450ºC which was 911.58 HV to 955.29 HV. Based on the test results show that the deeper the surface of the hardness value decreases until it reaches the same hardness as the initial material of the SKD61 specimen. The optimum hardness value is reached at 0.05 mm or 5µm depth. This indicates that the deeper the concentration of nitrogen at a lower concentration and the possibility of the occurrence of metal nitride compounds, the fewer the resulting violence decreases. On the surface edge of the first tracking sample it obtained hardness of 955.29 kgf/mm^2, while at a depth of 1.00 mm (core) the hardness value obtained was 485.42 kgf/mm^2. This indicates that the resulting hardness change in the nitriding process temperature of 500ºC covers a depth of 1mm until the resulting hardness is the same as the initial hardness of the material before hardening, which is 460 HV at a depth of 2 mm from the metal surface.

3.2.6. SKD61 Hardness after Plasma Nitriding at 550ºC.
Hardness testing with the Vickers technique for the SKD61 test sample after the plasma nitriding process was carried out by tracking the depth from 0.02 mm to the tracking depth of 0.4 mm from the sample surface and 100 g tracking load as shown in Figure 8.

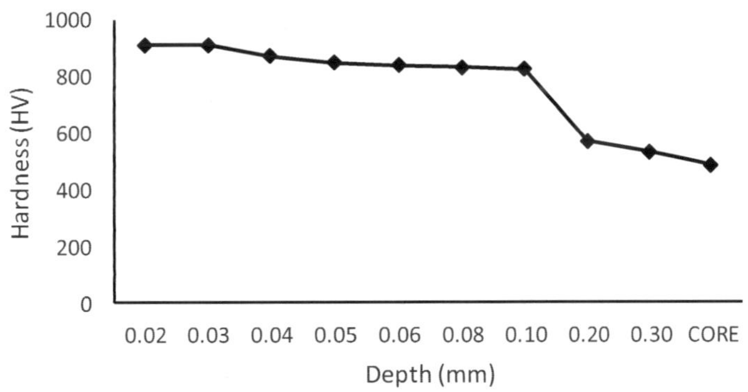

Figure 8. Hardness of the specimen after 10-hour plasma nitriding at 550ºC

Based on the results of the hardness test, it can be seen the hardness value of each tracking. Namely: 1) 1st tracking at 0.02mm, 2) 2nd tracking at 0.03mm, 3) 3rd tracking at 0.04mm, 4) 4th tracking at 0.05mm, 5) tracking 5th at 0.06 mm, 6) 6th tracking at 0.08 mm, 7) 7th tracking at 0.10 mm, 8) 8th tracking at 0.20 mm, 9) 9th tracking at 0.30 mm, and tracking above 1.0 mm is the 10th tracking (core / core section). The optimum hardness value is reached at a depth of 0.1 mm or 100µm. This indicates that the nitrogen diffusion zone is getting better inside Based on the results of the measurement of the hardness shown in the graph above, the measurement results of SKD61 material after the plasma nitriding process at a temperature of 550ºC decreased from the previous hardness at nitriding temperature of 500ºC which was 955.29 HV to 911.58 HV. However, if we see the spread of nitriding hardness of 550ºC more evenly than nitriding at a temperature of 500ºC, this can be seen from the difference in hardness value at 0.3mm depth on the nitriding test specimens of 550ºC higher than the hardness at nitriding temperature 500ºC with the depth of hardness same. While the decrease in hardness on the surface of nitriding samples at a temperature of 550ºC is due to an imbalance between the rate of deposition to the rate of diffusion or the reaction of nitrogen atoms (N) so that accumulation occurs. Deposition rate is influenced by nitriding pressure while deposition rate is influenced by substrate temperature [1].

Based on the test results show that the deeper the surface of the hardness value decreases until it reaches the same hardness as the initial material of the SKD61 Based on the test results show that the deeper the surface of the hardness value decreases until it reaches the same hardness as the initial material of the SKD61 specimen. The optimum hardness value is reached at a depth of 0.1 mm or 100μm. This indicates that the nitrogen diffusion zone is getting better into the surface to produce the optimum hardness value. At the surface edge of the first tracking sample it obtained a hardness of 911.58 kg f /mm^2, while at a depth of 1.00 mm (core) the hardness value obtained was 485.42 kg f /mm^2. This indicates that the resulting hardness change in the nitriding process temperature of 550ºC covers up to a depth of 1mm until the resulting hardness is the same as the initial hardness of the material before hardening which is 460 HV at a depth of 2 mm from the metal surface. From the results of measurements and testing of hardness of SKD61 specimen material after the temperature nitriding process of 350ºC, 400ºC, 450ºC, 500ºC, and 550ºC a data analysis can be taken as shown in Table 1.

Table 1. Distribution of Hardness of plasma nitriding results at 350ºC, 400ºC, 450ºC, 500ºC, and 550ºC for each at 10 hours

Depth (mm)	Hardness Vickers at Nitriding Temperature				
	350 °C	400 °C	450 °C	500 °C	550 °C
0.02	818.20	776.70	911.58	955.29	911.58
0.03	783.41	726.20	911.58	955.29	911.58
0.04	579.11	720.30	870.79	946.30	870.79
0.05	565.55	675.10	803.98	903.19	847.63
0.06	516.34	634.00	750.84	894.93	840.11
0.08	498.81	624.30	638.94	702.80	832.69
0.10	488.72	601.10	619.58	638.94	825.37
0.20	472.56	570.70	570.66	531.04	570.66
0.30	472.56	565.50	531.04	509.22	531.40
0.40	472.56	485.40	485.42	485.42	485.42

From figures above, it can be seen that the optimum hardness value is achieved when the hardness of 0.1 mm or 100 μm with the lowest hardness is reached at 350ºC and the highest hardness is reached when the nitriding temperature is 550ºC.

3.2.7. EDS (Energy Dispersive Spectroscopy) Testing.

The sample was cut using diamond saw so as not to damage the surface profile. In this study the SEM tool used is the JEOL type JSM-6390A brand which is equipped with EDS (Energy dispersive X-Ray Spectroscopy). EDS is used to determine the amount of concentration of element N which diffuses into the surface of the material. At this stage the position of observation is taken at the cross section of the sample surface. The results of this test are shown in Table 2.

Table 2. Qualitative analysis of plasma nitriding for 10 hours of SKD61 specimens.

Element	Mass Percentage (%) for temperature process of DC Plasma Nitriding				
	350 °C	400 °C	450 °C	500 °C	550 °C
N*	0.36	1.19	1.28	1.28	1.75
O	4.33	-	4.34	4.34	-
V	1.04	1.14	0.42	0.42	-
Cr	6.54	4.13	5.03	5.03	4.60
Fe	87.72	93.36	88.92	88.92	93.65

From the results were obtained that after nitriding at 350ºC has a nitrogen content of 0.36% mass. The appearance of O atoms in the EDS data is caused by the presence of oxygen trapped inside the nitriding chamber during the nitriding process. Other compounds that may be formed are Cr_2N, and VN so on EDS data elements Cr and V are detected [12, 16]. From the test results it was found that for the nitrided area after being at 400ºC it had nitrogen content of 1.19% mass. From the test results it was found that after being nitrided at 450ºC it had nitrogen content of 1.28% mass. From the test results it was found that after being nitrided at a temperature of 500ºC it had a 0nitrogen content of 1.54% mass. From the test results it was found that after nitriding at a temperature of 550ºC it had a nitrogen content of 1.75% mass. Seeing from the results of testing the qualitative analysis of plasma nitriding temperature 350ºC, 400ºC, 450ºC, 500ºC, and 550ºC for 10 hours that the influence of temperature affects the amount of diffuse nitrogen concentration or deposited on the surface of SKD61 material test specimens. The higher the nitriding temperature of plasma, the higher the nitrogen content obtained. The nitrogen element that exists when compared to the FeN diagram [17] is likely to form compounds Fe_2N, Fe_3N, Fe_4N, AlN, VN, Cr_2N. The properties of the nitriding compounds are very hard. However, for SKD61 material types classified as low carbon steel (<0.8% C) the ideal nitrogen element content is 11.2% mass so that Fe_2N compounds that have very hard properties can be obtained. The effect of plasma nitriding temperature on nitrogen concentrations obtained in SKD61 steel is shown in Tables 3 and Figure 9.

Table 3. Effect of nitriding temperature on nitrogen concentration obtained on SKD61 steel

Nitridation Temperature (°C)	Nitrogen Concentration (Mass%)
350	0.36
400	1.19
450	1.28
500	1.54
550	1.75

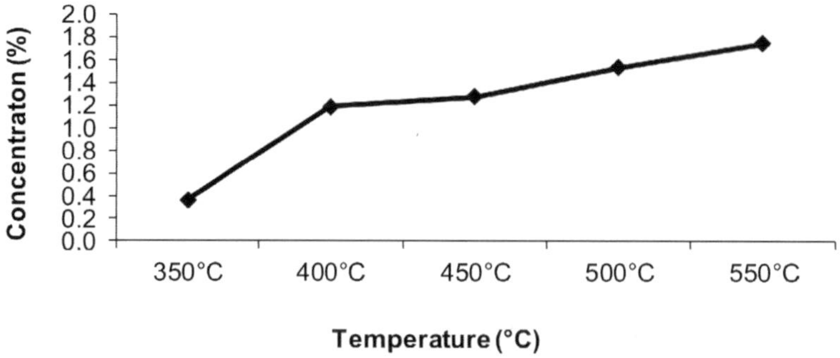

Figure 9. Effect of Nitriding Temperature on Nitrogen Concentration in SKD61 Steel.

3.2.8. Metallographic testing with Scanning Electron Microscope.
Metallographic tests are needed to determine the microstructure of SKD61 material especially after the nitriding process. This test was carried out using SEM whose results are shown in Figure 10.a for nitriding temperature of 350°C, Figure 10.b for nitriding temperature of 400°C, Figure 10.c for nitriding temperature of 450°C, Figure 10.d for nitriding temperature of 500°C, and Figure 10.e for nitriding temperature of 550°C

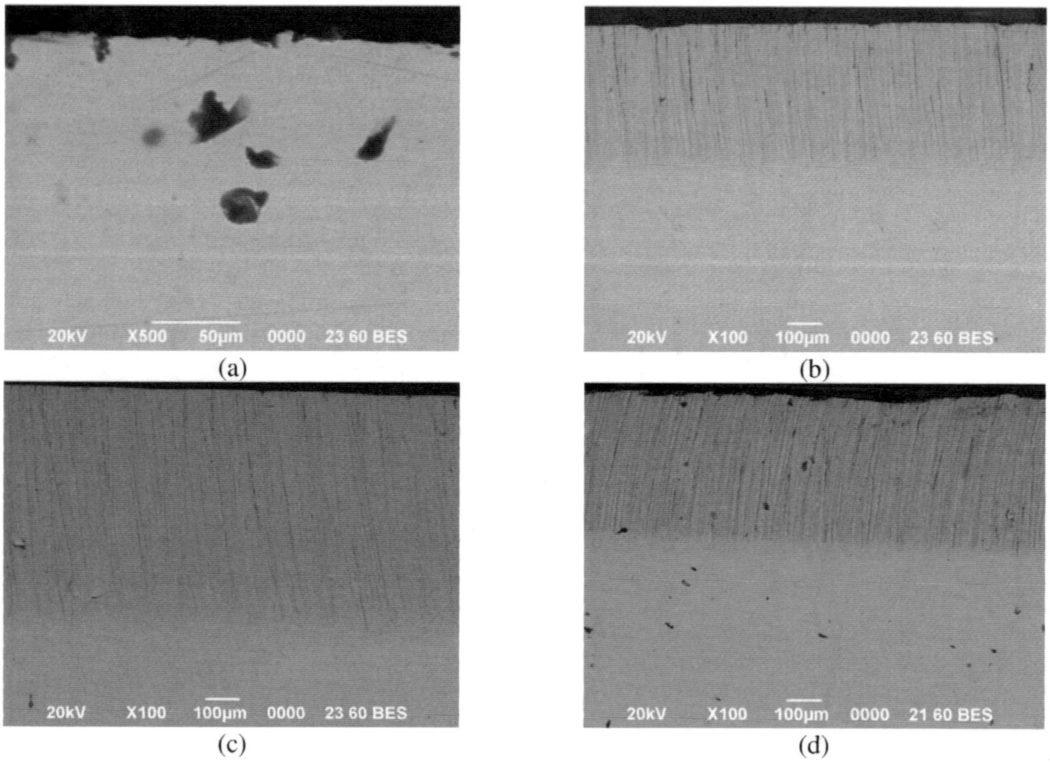

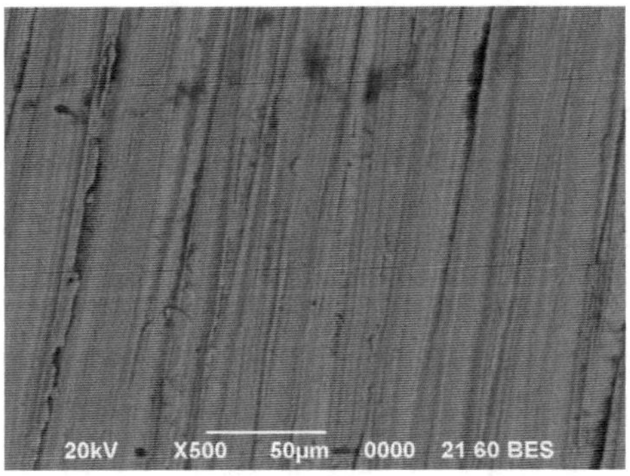

(e)

Figure 10. Microstructure Nitriding plasma at (a) 350ºC, (b) 400ºC, (c) 450ºC , (d) 500ºC, and (e) 550ºC, process for 10 hours

From the observation of microstructure using SEM the hard layer of Fe_4N and Fe_2N nitride compounds cannot be clearly seen, this is due to the nitrogen concentration obtained 0.36% N. If compared to the Fe-N diagram the nitrogen content below 1% is likely to dissolve in the Fe atom which forms the δFe phase [15]. The phase formed by delta iron (δFe) is a phase of the diffusion zone in which nitrogen atoms dissolve interstitially in ferrite matrix [17]. In Figure 10.b, SEM observations can be seen that the hard layer of nitride is formed, possibly the hard nitride formed is FeN, δFe, but the nitrogen element is still soluble in the Fe atom [15]. From the results of SEM nitriding observations on 450ºC temperature specimens, it was very clearly seen the hard nitride layer, when compared to the Fe-N diagram, the possible phase formed was Fe because the nitrogen elements obtained are 1.28%, however this type of phase was not harder than the phase Fe_2N, Fe_4N, and ε phase [15]. From the results of SEM nitriding observations on 500ºC temperature specimens, it is clear that the nitride hard layer, when compared to the Fe-N diagram, is likely to be formed γFe because the nitrogen elements obtained are 1.75% [15], however the phase concentrations obtained are still more a lot compared to the nitriding result of 450ºC . Apart from the δFe phase, other hard nitrides may also be formed such as Cr_2N, VN.

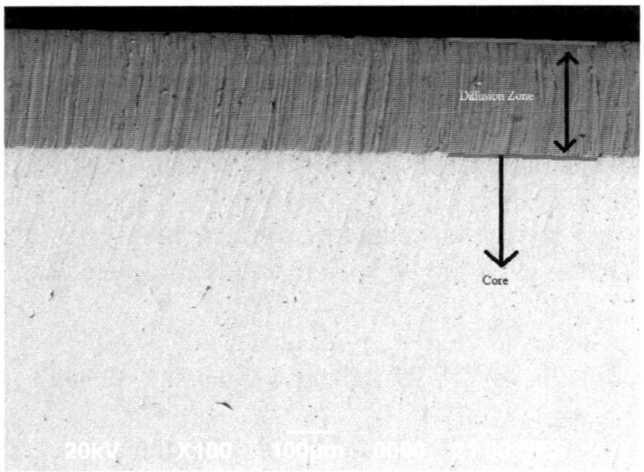

Figure 11. View of diffusion zone of Nitriding at 550ºC plasma for 10 hours

From the results of SEM nitriding observations in 550ºC temperature specimens, it is very clear that the nitride hard layer, when compared to the Fe-N diagram, is likely to form δFe because the nitrogen element obtained is 2.68% mass [15], however the phase γFe concentration obtained is still more than the nitriding result of 500ºC. Other compounds formed are like VN, Cr2N if you look at the EDS data obtained.

From Figure 11, the results of the nitriding SEM observations at a temperature of 550ºC with a magnification of 100 X, it can be seen that the diffusion zone is the phases where nitrogen elements dissolve in the Fe atom. According to ASM vol 4 p. 946, the nitride layer formed from nitriding results with nitrogen concentrations below 5% is a diffusion zone, while compound layer zones are zones where hard nitride phases such as Fe4N, Fe2N are formed when the nitrogen concentration is above 5% mass [15].

Overall, if you see the results of SEM nitriding of SKD61 specimens you can conclude that nitridation at 350ºC produces the amount the lowest nitrogen compared to nitriding temperatures of 400ºC, 450ºC, 500ºC, and 550ºC. The effect of temperature also determines the diffusion of nitrogen that occurs, resulting in hard nitride phases such as Fe-N, γ-Fe, α-Fe, δ-Fe. According to ASM volume 4 p 946 [19, 20] to obtain hard phases such as Fe4N, and Fe2–3N during the nitriding process, it is necessary to pay attention to the equilibrium between the gas mixture N2 and H2. The hard structure of the nitride Fe4N layer is obtained when nitrogen gas flowing in the nitriding chamber is 15-30% nitrogen, and Fe2–3N nitride can be obtained by flowing 60 - 70% nitrogen with a mixture of C2H4 (methane) 1 - 3% [11-20]. The recapitulation of the test results and observations is shown in Table 4 as follows:

Table 4. Recapitulation of Vickers test results and SEM-EDS observations

Item	Before Hardening	After Hardening	Temperature process DC plasma nitridation (°C)				
			350	400	450	500	550
Hardness (HV)	460	531.32	488.7	601.1	619.6	638.9	825.4
Depth (mm)	2	2	0.1	0.1	0.1	0.1	0.1
Microstructure	Ferrite-pearlite	martensite	δFe	FeN, δFe	FeN, γFe	FeN, Cr_2N	VN, $Cr_{,2}N$
Nitrogen Concentration (%)	0	0	0.36	1.19	1.28	1.54	1.75

4. Conclusion

1. The best formation of hard nitride layer is when plasma nitriding temperature of 550ºC for 10 hours with the nitrogen concentration obtained is 1.75% mass and the possibility of the phase formed is δFe, FeN, Cr2N, and VN.
2. Testing of Vickers nitriding plasma at a temperature of 500ºC for 10 hours has the highest hardness value compared to other nitriding specimens which is equal to VHN 955.29 kgf/mm^2 with a depth of 100 μm.
3. The optimum condition is when the hardness produced increases 1.79 times from the previous 531.32 kgf/mm^2 after hardening to 955.29 kgf /mm^2 after the nitriding process of plasma at a temperature of 500ºC.
4. Vickers hardness testing of SKD61 specimens after hardening increased 1.15 times from the previous 460 kgf/mm^2 to 531.32 kgf/mm^2.
5. The effect of Plasma Nitridation has been shown to improve the surface quality of SKD61 materials, especially surface hardness with optimum conditions achieved at plasma nitriding temperatures of 500 ° C for 10 hours.
6. Metallographic observation of SEM-EDS shows that the nitriding temperature affects the concentration and depth of diffusion of nitrogen in the SKD61 material.

5. References

[1] Sudjadi U 2013 Study About Surface Hardening on Local Disc Brakes with Direct Current Plasma Nitrocarburizing Apparatus *Adv. Mater. Res.* **789** 383-6
[2] Leonard Devila G et al 2008 *Physics Scripta* T131 014011.
[3] Dutka V A 2008 *J.* Superhard *Mater.* **30** 338–348
[4] Makishi et al 2009 *Eng. Min. J.*
[5] Gredelj S et al 2006 *Thin Solid Films*
[6] Valencia R et al 2004 *Braz. J. Phys.* **34**
[7] Sudjadi U 2012 *Development of RF-Plasma Nitriding Tools into RF-Plasma Nitrocarburizing and Surface Hardening Materials in Hotcell and NPP Components research* (Indonesia Minister of Research and Tech. Office, Indonesia)
[8] Sudjadi U et al 2014 Characterization of local bolt (C=0.071%) after surface hardened at a temperature of 400^0 C for 7 hours with RF plasma nitrocarburizing apparatus *Appl. Mech. Mater.* **575** 446-51.
[9] Amstead B H et al 1995 *Mechanical Technology* Vol I 7th edition (Jakarta, Indonesia, Erlangga)
[10] Smallman R E 1991, *Modern Physical Metallurgy* 5th edition (Jakarta, Indonesia: Gramedia Main Library)
[11] Schonmetz, Alois and Karl Gruber, 1985 *Material Knowledge in Work Metal* (Bandung, Indonesia: Space First Press)
[12] Van V and Lawrence 1985 *Elements of Materials Science and Engineering*
[13] Koswara and Engkos 1999 *Metal Testing* (Bandung, Indonesia: Humanities First Press)
[14] William D C Jr and David G. R 2018 *Material Sciene and Engineering an Introduction* vol 199
[15] Surdia, Tata, and Saito S. 2000 *Knowledge of Engineering Materials* (Jakarta, Indonesia: Pradnya Paramita)
[16] Totten G E, Bates C E and Clinson N A 1993, *Handbook of Quenchants and Quenching Technology* (USA ASM International)
[17] Smallman R E 1999 *Modern Physical Metallurgy* 5th edition (Jakarta, Indonesia: Gramedia Main Library)
[18] *ASM International* vol 3 Phase Diagram pp 842
[19] *ASM International* vol 4 pp 944
[20] Anonym 1988 *Metal Handbook, Heat Treating, Cleaning, and Finishing* vol 2 (ASTM United States)

Acknowledgement

The authors contributed equally to this work. The authors also would like to thank Dr. Jan Setiawan for reviewing and proofreading the manuscript.

IOP Conf. Series: Materials Science and Engineering **839** (2020) 012020 doi:10.1088/1757-899X/839/1/012020

The effect of pumice composition upon transmission loss and impact strength of pumice hybrid composite–coconut fiber/gypsum

N P G Suardana[1,*], C I P K Kencanawati [1], I K G Sugita[1], I G N Wardana[2] and I P G R Putra[1]

[1] Material Laboratory, Mechanical Engineering Udayana University, Bali, Indonesia
[2] Engineering Geology Laboratory, Civil Engineering Udayana University, Bali, Indonesia
[*] Corresponding Author: npgsuardana@unud.ac.id

Abstract. Voice could not be separated from ears, but there is other sound called noise. Soundproofing panel in a room is highly important to hold and absorb sound in order to prevent sound to come and out of the room so the room will be comfortable. The objective of this study is to find the characteristic of transmission loss and the value of impact strength caused by the effect of pumice and gypsum composition. This study used 2 mm pumice combined with coconut fiber in 20% of fiber volume fraction, by using gypsum and glue matrix. The varied composition of this study used two comparisons Gypsum: Pumice: Glue of 1:1:1/2, and 1:2:1/2. Transmission loss was tested by using test equipment (ASTM-E-1050) and Sound Level Master (SL-4011), while the impact strength was tested by using Charpy Impact test equipment (ISO179-2). Result of this study showed that the lowest transmission loss occurred in specimen with composition 1:1:1/2 with value 13.48dB, while the highest transmission loss occurred in specimen with composition 1:2:1/2 with value 25.21dB. The lowest impact strength occurred in specimen with composition 1:2:1/2 with value 4.293kJ/m2, while the highest mean value for impact strength was 5,534 kJ/m2 which occurred in composition 1:1:1/2.

1. Introduction

By the advanced development of technology, the human tools development is improving as well. The variations are in form of information, communication, production, transportation or entertainment equipment. Most of the equipment produce unwanted sounds that cause noises. To have comfortable room which is far away from noises becomes everyone desire. Controlling is done to block sounds in room so the sound intensity becomes weak. Therefore, the room should have wall cover to muffle sounds. The existing silencer are the porous material, resonator, and panel [1]. Some researches have been done before about silencer by using different materials such as: developing silencer material from bamboo fiber as good as glasswool [2], developing silencer from recycled polyester fiber.

Until today, coconut fiber is only developed to be many industrial products such as carpet, mat and other functional handicraft. Few people know that coconut fiber can be used as reinforcement in composite. Processing coconut fiber become effective as new engineering material are producing nature-friendly new composite, cheaper with high economical value. Pumice is one of materials originated from volcano eruption. Then it has natural cooling process and sediment in the subsoil for years. Pumice and pearlite can be used as main material to make light concrete, because it has some characteristics such as high porosity, low density, high thermal insulation, and shock resistant as earthquake. So, there should be improvement by changing conventional concrete material (cement) with a polymer composite material [3].

Gypsum ($CaSO_4.2H_20$) is a chemical compound arranged from calcium, sulfate group and water. Gypsum is not heavy and fire proof, with density 2.32g/cm^3. Gypsum is chosen as silencer composite material because it has characteristic of noise reduction and good sound absorber. The alternative material that can be used as absorber panel based on pumice hybrid as filler and combined with coconut fiber volume of 20% and then bounded by gypsum and glue as matrix upon loss transmission loss and impact strength. It is expected from this research, there will be new innovation in technology development of composite material reinforced with natural fiber and pumice stone in the application as sound absorber .

2. Research Method

2.1 Process of Making Pumice Particle and Coconut Fiber Treatment

Pumice was cleaned to remove dust and dirt by using clean water by rubbing with brush to get cleaner and fast result. After that, pumice was rinsed by aquadest and then dried by using oven on 70° C for 24 hours. The dried pumice was pounded to be in smaller size (±2mm) by sieving machine.

The process of coconut fiber treatment was done through the following some steps. Firstly, to remove impurities, the coconut fiber was washed until clean. Secondly, coconut fiber was boiled on 100°C for 15 minutes to reduce the impurities attached on the coconut fiber, then it was dried for 24 hours in oven on 70° C. Thirdly, the fiber was cut into 10mm, and treated with NaOH with concentration 5% for 2 hours and then it was washed so it was really clean. Finally, the fiber was dried again on 70°C for 24 hours.

2.2 The Process of Making Composite Mold

Ceramic mold in square shaped 400 x 400 mm was prepared for the composite pedestal and cover. Mold dimension 300 x 300 mm with 10 mm thickness was prepared as well. Before molding, ceramic surfaces were cleaned with clean water to reduce impurities factor. The mold was placed based on composite thickness that was 10 mm and glued. The mold was ready to process molding as shown by Figure 1.

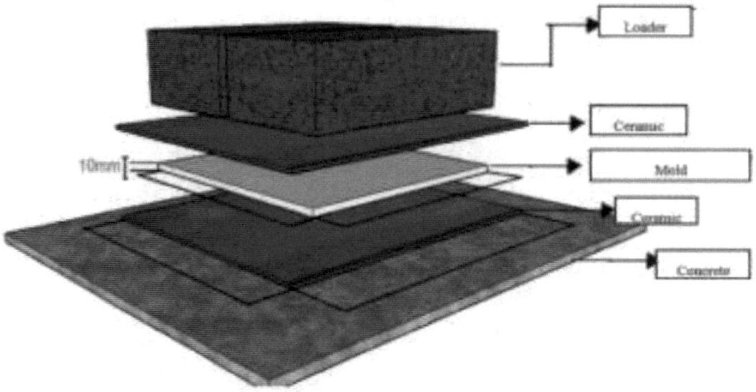

Figure 1. The series of Composite Mold.

Table 1. Transmission Loss and Impact Strength values.

Composition of Gypsum:Pumice: Glue	Fiber volume fraction (%)	Mass of fiber (gram)	Gypsum volume (cm^3)	Gypsum mass (gram)	Volume of pumice (cm^3)	Mass of pumice (gram)	Glue volume (cm^3)	Glue mass (gram)
(1:1:1/2)	20	226,8	288	668,16	288	184,60	144	241,92
(1:2:1/2)	20	226,8	206	477,92	411	263,45	103	173,04

2.3 Composite Molding and Post Curing Process

Gypsum and glue were mixed and stirred until evenly distributed. The pumice that was filtered by using 2 mm sieve was poured with gypsum and glue mixture, and then stirred until evenly distributed. Coconut fiber was added so it became hybrid composites into container based on the compositions as mentioned in Table 1. The mold was smeared with glycerin evenly and thinly. The mixtures were poured into the mold and close the upper mold cover slowly. If there are some voids, direct the trapped voids by pressing and then provide loading which possibly pressed it thoroughly for 24 hours. The other specimen variance could be done by repeating the processes from the beginning. After being dried, the composite was taken from the mold by using a tool slowly and carefully. The flat of released composites was checked by using flat glass and then the composite was ready to be treated, cut, and tested. Process of treatment by post curing process was done by putting the composite into oven on 70° C for 2 hours. It is intended to remove bubbles and steam trapped in the composite and to know whether the composite was homogeneous or not, i.e. the composite sheet did not bend. Every half of hour, composite sheet should be flipped to spread the heat on the sheet surface which was intended to avoid the bending of the sheet.

2.4 Method of Transmission Loss Test

Testing method was done by placing acoustic panel on the test tool in which acoustic panel would be placed between earpiece-in and earpiece-out. The sound sources were in frequency of input 125Hz, 250Hz, 500Hz, 1000Hz, 1500Hz, 2000Hz, 2500Hz, 3000Hz, 3500Hz and 4000Hz. The further step was data collection in earpiece-in by sound level meter for 150 seconds, with data collection interval of 5 seconds and continued to collect data in earpiece-out with the same way.

IOP Conf. Series: Materials Science and Engineering **839** (2020) 012020 doi:10.1088/1757-899X/839/1/012020

3. Result and Discussion

3.1 The Result of Transmission Loss Test

The transmission loss test of hybrid composite of pumice-coconut fiber/gypsum was calculated using formula so the final result was in form of transmission loss material value by using the following formula:

$$TL = I_0 - I \quad (1)$$

Description:

- TL = Transmission Loss (dB)
- I_0 = Sound In (dB)
- I = Sound Out (dB)

The mean value data of transmission loss test of composite can be seen in the following table:

Table 2. The Mean value of Transmission Loss (dB).

Composition of Composite with Fiber Volume Fraction 20%	Transmission Loss (dB)
1:1:1/2	13.48
1:2:1/2	25.21

From the above result, the data were plotted into bar chart to describe the mean value of transmission loss of the specimen.

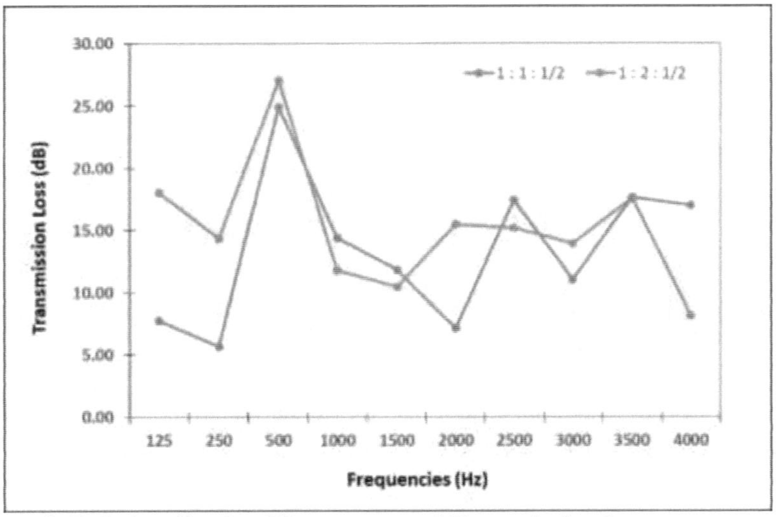

Figure 2. Relation Graph of Frequency and Transmission Loss on composite.

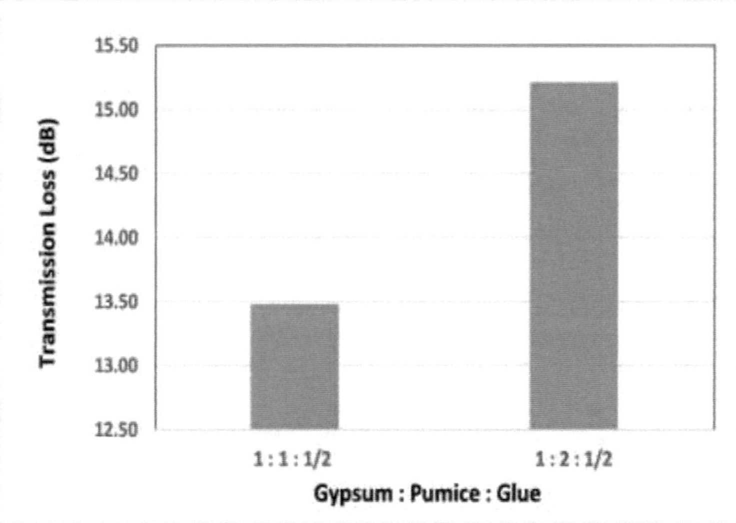

Figure 3. Bar chart of transmission loss of hybrid composite with various composition.

The above Figure shows the result of TL test for pumice hybrid composite with various composition consisting of Gypsum, Pumice, and Glue (1:1:1/2, and 1:2:1/2.) with fiber volume fraction 20%. The results showed that in low frequency of 125 Hz, the TL value was the lowest, then in frequency 500Hz, it tended to be the highest TL and in frequency 1000Hz started to decrease and fluctuate until reached frequency 4000Hz (Figure 2). The highest average TL value occurred on the composite with composition 1:2:1/2 (Figure 3). It was due to the composition 1:2:1/2 contained more pumice, in which pumice has cavities and porous so it has potency to muffle sound well to be used as sound absorbent material. Sound absorption is mostly due to acoustic energy dissipation resulting from the porous material, porosity has been cited as the main factors that affect sound absorption [4,5,6] which can be seen in Figure 4 and Figure 5.

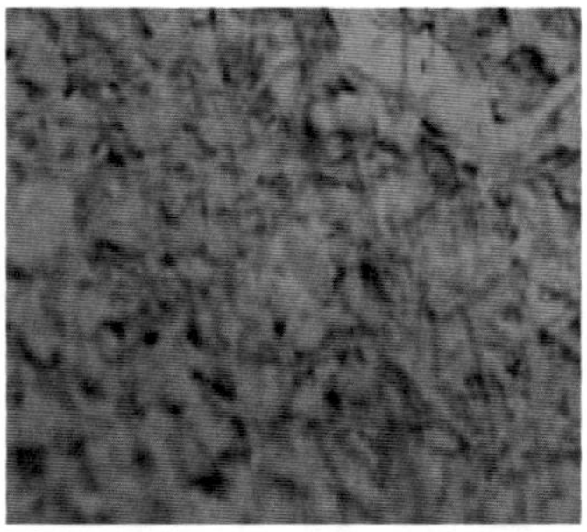

Figure 4. It can be seen that composition composite 1:2:1/2 shows more porous and the surface of specimen is fragile or not real.

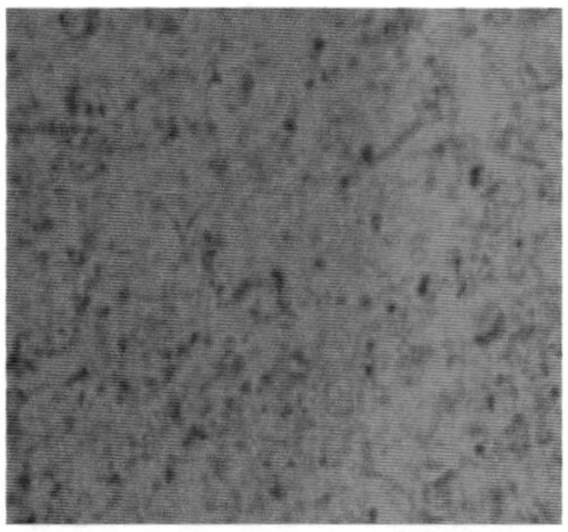

Figure 5. It can be seen that composition composite 1:1:1/2 shows evenly surface of the specimen.

From this specimen test data, the content of pumice in composite as well as gypsum comparison also affected the improvement of transmission loss in high frequency, and from these 2 material or specimen compositions that have been tested in all frequencies, it can be seen in Figure 2. It was found that the highest mean score of the transmission loss was in volume fraction 20% with comparison (1:2:1/2) with transmission loss 25.21 dB, while the lowest transmission loss was 13.48 dB with comparison (1:1:1/2).

3.2 The Result of Impact Test

The result of impact test upon the composite specimen to know the impact strength energy was described by the following Table and Figure:
Calculating the Impact Strength:

$$Is \frac{\Delta E}{A} \tag{2}$$

Description:

Is = Impact Strength (Nm/mm^2)
ΔE = Impact Energy (Nm)
A = Cross-sectional area (mm^2)

3.2.1 Discussion of Impact Test.

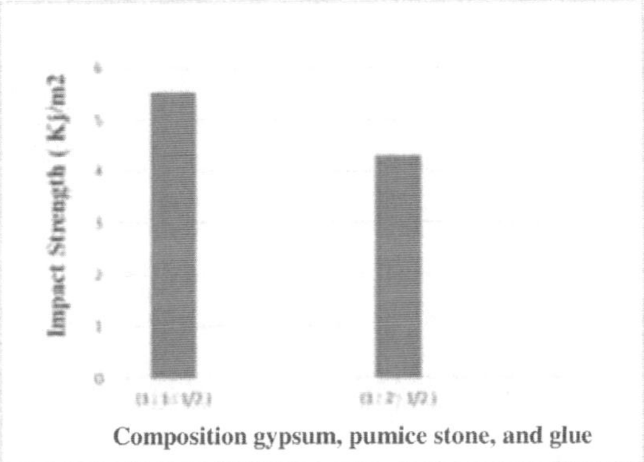

Figure 6. The Graph of Impact Strength on Hybrid Composite Specimen of the mixture of Gypsum, Pumice and Glue (1:1:1/2), (1:2:1/2), fiber volume fraction 20%.

Based on the calculation, the result of impact test can be seen in the following Figure 6. From the result of impact strength on hybrid composite of pumice and coconut fiber with fiber volume fraction 20% with matrix gypsum, it was found that the highest impact strength improvement was in composition 1:1:1/2) with the value of impact strength 5.534 kJ/m^2, in composition (1:2:1/2) the impact strength was 4.293 kJ/m^2 (Figure 6). The improvement of impact strength on fiber volume fraction 20% with composition (1:1:1/2) was caused by the appropriate combination of decreasing the number of pumice, so the gypsum and glue could bind the whole fiber and pumice optimally, and it increased the impact strength.

The decrease of impact strength on composition (1:2:1/2) because the content of gypsum in this composition was lesser than pumice, so this composition composite specimen was more fragile and uneven, in this composition, the gypsum matrix could not bind all of the pumice and fiber. It is also due to the characteristic of pumice more porous so it lowers the impact strength. This results is also agree with research [5] of pumice stone which concluded that the increasing of pumice stone can decrease the impact strength of material, which can be seen in Figure 7 and Figure 8.

IOP Conf. Series: Materials Science and Engineering **839** (2020) 012020 doi:10.1088/1757-899X/839/1/012020

The gypsum characteristic of which easily hardening and the level of hardening is high, so the addition of gypsum impacted the level of impact strength *(Is)*. The brake of gypsum specimen includes broken damage and damage that only causes cracking. For the gypsum specimen which was not strengthen by fiber, the impact strength was lower than the gypsum specimen with fiber [7].

Figure 7. The fracture of impact specimen with composition 1:1: ½ shows the gypsum matrix binds pumice and coconut fiber well so the density was very good.

Figure 8. The fracture of impact specimen with composition 1 : 2 : 1/2 shows more pores and the surface is fragile or uneven because in this composition, gypsum matrix could not bind the entire pumice and fiber well.

3.3 SEM

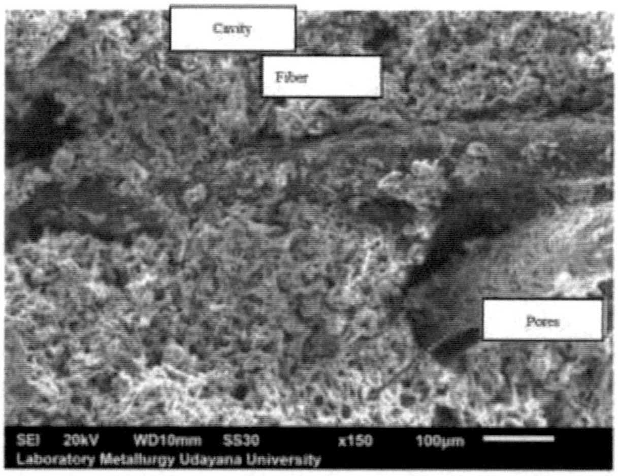

Figure 9. SEM photo of hybrid composite of pumice with composition (1:1:$^{1}/_{2}$) (Magnification 150x).

SEM test was done to know the characteristic of composite specimen and see with 5.534 kJ/m^2 clearly the existing mixture on the specimen. In the Figure 9, can be seen that density between gypsum matrix and pumice particle is very tight, and the coconut fiber attached well, so it made this composition became strongest.

Figure 10. SEM photo of hybrid composite of pumice in composition (1:2:1/2) (Magnification 150x).

Figure 10 shows SEM photo by using magnification 150x. In the above Figure, it can be seen that there are more cavities because the gypsum and glue could not bind entirely the fiber and pumice, so this composition is more fragile from the other composites.

4. Conclusion

Based on the data analysis and discussion, it can be summarized that:
The lowest transmission loss occurred in fiber volume fraction specimen 20% in composition 1:1:1/2 with mean value 13.48 dB. The highest transmission loss with mean value 25.21 dB occurred in comparison composition 1:2:1/2. The lowest impact strength occurred in fiber volume fraction specimen 20% with impact strength mean value 4.293 kJ/m^2 in comparison 1:2:1/2, while the highest impact strength occurred in comparison 1:1:1/2 with value 5.534 kJ/m^2.

5. References

[1] Lee Y and Changwhan Joo 2003 Sound absorbtion properties of recycled polyester fibrous assembly absorbers *J. AUTEX Research* **3**
[2] Koizumi T, Tsujiuchi N and Adachi A 2002 The development of sound absorbing materials using natural bamboo fibers *Paper from: High Performance Structures and Composites.* ISBN 1-85312-904-6.
[3] Karo Karo P 2013 Perbandingan hasil uji fisis komposit polimer berbasis perlit dengan batu apung *J. Fisika FMIPA Universitas Lampung* **01** No 02
[4] Fauzi, Tamrin, Sembiring A D and Sani R A 2016 Characteristics of polymer concrete from pumice stone and rubber thread waste with polyurethane as natural bonding *Am. J. Phys. Chem.* **5** 26-34
[5] Yilmaz N, Michielsen S and Michielsen S 2011 Effects of porosity, fiber size and layering sequence on sound absorption performance of needle-punched nonwovens *J. Appl. Polym. Sci.* **121** 3056-69
[6] Suardana N P G, Lokantara I P and Sugita I K G 2015 Panel akustik ramah lingkungan berbahan dasar limbah batu apung dengan pengikat poliester *Proceding Seminar Nasional Tahunan Teknik Mesin* XIV (SNTTM XIV).

[7] Ramezani H, Shahdab S and Nouri A 2012 Study on effects of wood fiber content on physical, mechanical, and acoustical properties of wood-fiber-filled gypsum composites *J. Mater. Res.* **15** 236-41

Acknowledgment
This research was funded by *DRPM* of Research Strengthening and Development of Directorate General of Research, Technology, and Higher Education Ministry based on Research Contract No: 171.93/UN14.4.A/LT/2018, dated 19 February 2018.

Mixed convection of single-walled carbon nanotubes in a triangular cavity containing a pentagonal impediment

Waqar A Khan[1*] and Zafar H Khan [2]

[1]Department of Mechanical Engineering, College of Engineering, Prince Mohammad Bin Fahd University, Al Khobar 31952, Kingdom of Saudi Arabia
[2]State Key Laboratory of Hydraulics and Mountain River Engineering, College of Water Resource and Hydropower, Sichuan University, Chengdu 610065, P. R. China
wkhan@pmu.edu.sa

Abstract. Mixed convection of nanofluid (a homogeneous mixture of water and single-walled carbon nanotubes) is investigated numerically in a triangular cavity containing an adiabatic pentagonal obstacle close to the top surface. Analytical models are employed for effective thermophysical properties of nanofluid regarding carbon nanotubes and water. The top wall of the cavity is kept at a higher temperature, while the inclined sidewalls are kept at a lower temperature. A constant magnetic field of intensity B_0 is introduced in the momentum equation. The effects of radiation and viscous heating on heat transfer are neglected. The impacts of Richardson number on the dimensionless velocity and temperature are presented graphically in terms of streamlines and isotherms. The Galerkin finite element method with a penalty function is employed to obtain the solution of dimensionless partial differential equations. The numerical results are examined for governing parameters, including volume fraction of carbon nanotubes, Richardson, and Hartmann numbers. It is concluded that the Nusselt number increases with increasing the governing parameters in the presence of an adiabatic pentagonal obstacle.

Keywords: Mixed convection; Magnetic field; SWCNT-water nanofluid; Triangular cavity; pentagonal block; Finite element method.

1. Introduction

The rate of heat transfer can be increased by increasing the thermal conductivity of base fluids. Due to high thermal conductivity, CNTs are used to obtain high heat transfer rates in several engineering applications like cooling of electronic components, furnaces, thermal storage systems, high-performance building insulation, drying technologies, and heat exchangers. In such applications, the transfer of heat energy plays an important role. Several studies exist related to the heat convection in different cavity configurations filled with various fluids and imposed boundary conditions on the walls. For instance, Selimefendigil [1] and Jafari et al. [2] studied mixed convection of single- and multiple-walled carbon nanotubes in cavities and noticed a better improvement in the heat transfer than other nanoparticles. The effects of mixed convection in a wavy walled lid-driven enclosure filled with nanofluid were also investigated by Öztop et al. [3]. They discovered an increase/decrease in

Nusselt number with the increasing volume fraction of nanoparticles based on governing parameters. Selimefendigil and Öztop [4] employed various thermal boundary conditions and studied mixed convection of a nanofluid in a lid-driven cavity. They concluded that the solid volume fraction of nanoparticles enhances the rate of heat transfer in the cavity. Later, Selimefendigil and Öztop [5] extended their study to a moving wall and internal heat generation. Using non-uniform heating conditions on both sidewalls, Sivasankaran et al. [6] deliberated mixed convection in a lid-driven cavity. It was demonstrated that the Nusselt number increases with increasing amplitude ratio. Mekroussi et al. [7] also found higher heat transfer rates in the cavity due to mixed convection and proved that the wavy walls have a powerful impact on enhancing heat transfer.

Sivasankaran et al. [8] studied mixed convection in a lid-driven square cavity with a sinusoidal boundary condition at the base surface. Their results depend upon Richardson and Hartmann numbers. Sivakumar and Sivasankaran [9] studied mixed convection numerically in an inclined square cavity. They noticed a substantial increase in the heat transfer rate in the cavity. Using extended-Darcy model, Sivasankaran and Pan [10] examined mixed convection in a cavity with nonuniform thermal boundary conditions. It was observed that an increase in amplitude ratio increases the heat transfer rate. Bakar et al. [11] examined the behavior of mixed convection numerically with internal heat generation/absorption in a square lid-driven cavity.

They observed the reduction of heat transfer with an increase in heat generation. For pseudoplastic fluids, Manchanda and Gangawane [12] reported an insignificant effect on fluid and thermal structure during mixed convection. Gangawane and Manikandan [13] employed isothermal and isoflux thermal boundary conditions for mixed convection of Newtonian fluids in a lid-driven cavity. They noticed a significant increase in the heat transfer with increasing Reynolds number up to 200 and then decreases. Gangawane [14] investigated mixed convective flow in a cavity using isoflux boundary condition on the walls. He confirmed the numerical results of [13]. Kalteh et al. [15] used a triangular heat source in a lid-driven square cavity filled with different nanofluids. They confirmed a substantial increase in Nusselt number with the volume fraction of nanoparticles.

Based on Galerkin Weighted Residual technique, Ali et al. [16] studied mixed convection in a hexagonal cavity in the presence of a magnetic field. They assumed isothermal boundary conditions and recovered the considerable impacts of governing parameters on the flow and the dimensionless heat transfer. Louaraychi et al. [17] investigated the effects of major parameters on the behavior of mixed convection in lid-driven cavities. They observed the zones showing the dominance of natural convection. A complete review of the mixed convection of nanofluids in several enclosures was presented by Izadi et al. [18]. It was concluded that the Nusselt number and pressure drop depends upon Reynolds and Richardson numbers as well as the volume fraction of nanoparticles. The impacts of pertinent parameters on the flow and heat transfer were examined by Ezzaraa et al. [19]. They found that the thermal performance of the cavity depends upon the emissivity of the walls, the velocity of the external flow, and the position of the outlet port.

Using water–Al_2O_3 nanofluid, Ghasemi and Aminossadati [20] conducted a numerical study on the mixed convection and compared two different scenarios of moving walls. They found a higher heat transfer rate for the downward sliding wall motion. Nayak et al. [21] considered a non-homogeneous model to examine the influence of governing parameters on the flow and heat transfer of a nanofluid in a partially heated enclosure. They noticed the effects of the governing parameters on both the flow and heat transfer of a nanofluid. Using Buongiorno's model and internal heat generation, Yu et al. [22] examined the behavior of nanofluid in an inclined lid-driven cavity. They observed the dependence of governing parameters on the streamlines, isotherms, and the local Nusselt number in the cavity. In the presence of a magnetic field, Javed and Siddiqui [23] investigated mixed convection in a square enclosure containing micropolar fluid. They noticed the dependence of mixed convective parameters in the convection regime. Kapil et al. [24] considered an enclosed cavity filled with nanofluid and inferred that the solid volume fraction of nanoparticles has a substantial impact on the heat transfer rate. Using numerical simulations in a cubical cavity, Al-Rashed et al. [25] studied mixed convection of nanofluids. They noted that the volume fraction of nanoparticles has a weak impact on heat transfer.

Using a square enclosure, Wang et al. [26] examined mixed convection and verified the dependence of heat transfer on the pulsating temperature of the inner cylinder. They concluded that the unsteady mixed convection enhances the heat transfer rate in the enclosure. Other studies related to mixed convection in lid-driven cavities filled in with Newtonian/no-Newtonian fluids under different thermal boundary conditions on the cavity walls can be found in [27-37]. Their findings confirm the results of the published work.

The above literature indicates the shortage of studies related to mixed convection in triangular cavities filled with CNTs and containing a pentagonal block. The main objective of this study is to investigate the effects of mixed convection parameter on streamlines, isotherms and heat transfer from a nanofluid comprising single-walled carbon nanotubes with water in a triangular cavity with an adiabatic pentagonal block.

2. Model Formulation

A 2-D triangular cavity, containing a heated pentagonal obstacle close to the top wall, is considered for the numerical analysis of mixed convection of the selected nanofluid. The top wall of the cavity is parallel to the x-axis of the Cartesian Coordinate system and is moving with velocity u_0 towards the right. The top wall is maintained at higher temperature T_h, whereas the inclined sidewalls are kept cold at lower temperature T_c, as shown in Figure 1. Consequently, the nanofluid motion is subjected to a buoyancy force owing to the differential heating and temperature difference. The gravitational force is acting vertically downward. A constant magnetic field of strength B_0 is enforced on the cavity walls. The magnetic Reynolds number is assumed to be negligible.

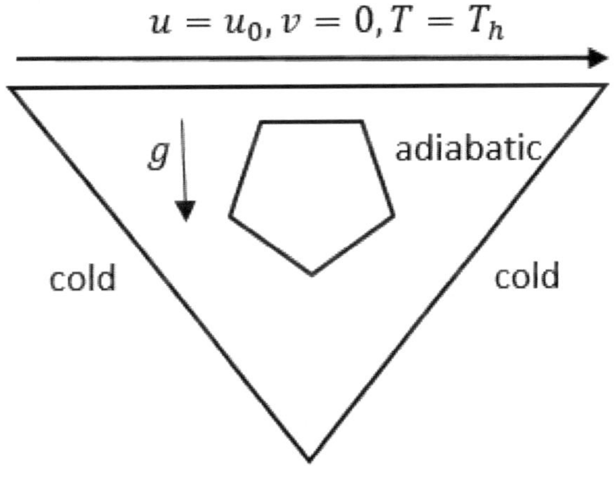

Figure 1. Geometry of triangular cavity with inner pentagonal obstacle.

Under these assumptions, the dimensionless governing equations can be written as [20]:

$$\frac{\partial U}{\partial X}+\frac{\partial V}{\partial Y}=0, \tag{1}$$

$$U\frac{\partial U}{\partial X}+V\frac{\partial U}{\partial Y}=-\frac{\rho_f}{\rho_{nf}}\frac{\partial P}{\partial X}+\frac{\upsilon_{nf}}{\upsilon_f}\frac{1}{\mathrm{Re}}\left(\frac{\partial^2 U}{\partial X^2}+\frac{\partial^2 U}{\partial Y^2}\right), \tag{2}$$

$$U\frac{\partial V}{\partial X}+V\frac{\partial V}{\partial Y}=-\frac{\rho_f}{\rho_{nf}}\frac{\partial P}{\partial Y}+\frac{\upsilon_{nf}}{\upsilon_f}\frac{1}{\mathrm{Re}}\left(\frac{\partial^2 V}{\partial X^2}+\frac{\partial^2 V}{\partial Y^2}\right)-\frac{\sigma_{nf}}{\sigma_f}\frac{Ha^2}{\mathrm{Re}}V+\frac{(\rho\beta)_{nf}}{(\rho\beta)_f}Ri\theta, \tag{3}$$

$$U\frac{\partial\theta}{\partial X}+V\frac{\partial\theta}{\partial Y}=\frac{1}{\Pr\operatorname{Re}}\frac{\alpha_{nf}}{\alpha_f}\left(\frac{\partial^2\theta}{\partial X^2}+\frac{\partial^2\theta}{\partial Y^2}\right), \tag{4}$$

The above dimensionless governing equations are obtained using the following dimensionless variables and parameters:

$$X=\frac{x}{L},\ Y=\frac{y}{L},\ U=\frac{u}{u_0},\ V=\frac{v}{u_0},\ P=\frac{p}{\rho_f u_0^2},\ \theta=\frac{T-T_c}{T_h-T_c},\ \operatorname{Re}=\frac{u_0 L}{v_f}$$

$$\Pr=\frac{\upsilon_f}{\alpha_f},\ Gr=\frac{\beta_f g\left(T_h-T_c\right)L^3}{\upsilon_f\alpha_f},\ Ha=B_0L\sqrt{\frac{\sigma_f}{\rho_f\upsilon_f}},\ Ri=\frac{Gr}{\operatorname{Re}^2}$$

where Pr, Gr, Ha, Ri are the Prandtl number (Pr ≈ 6.2 for water), Rayleigh number, Hartman number, and Richardson number, respectively. The corresponding boundary conditions in the non-dimensional form are given by

1. Top wall: $\begin{cases} U=1, \\ \theta=1, \quad 0\le X\le 1, Y=0.85 \\ V=0, \end{cases}$ (5)

2. Right inclined wall: $U=V=\theta=0$ (6)

3. Left inclined wall: $U=V=\theta=0$ (7)

4. For Inner pentagonal shaped obstacle temperature $\frac{\partial\theta}{\partial Y}=0$. (8)

5. Velocities at all walls of the obstacle are zero, i.e., $U=V=0$. (9)

The local and average Nusselt numbers along the domain of enclosure which is partially heated are described as:

$$Nu=-\frac{k_{nf}}{k_f}\frac{\partial\theta}{\partial Y}\bigg|_{Y=0.85},\ Nu_{avg}=\int_L Nu\,\mathrm{d}X. \tag{10}$$

The empirical relations for the effective physical properties for nanofluid concerning base fluid and nanoparticles are given as follows [38]:

$$\rho_{nf}=(1-\phi)\rho_f+\phi\rho_s,\ (\rho C_p)_{nf}=(1-\phi)(\rho C_p)_f+\phi(\rho C_p)_s,$$

$$\mu_{nf}=\frac{\mu_f}{\left(1-\phi\right)^{2.5}},\ \frac{k_{nf}}{k_f}=\frac{1-\phi+2\phi\frac{k_{CNT}}{k_{CNT}-k_f}\ln\frac{k_{CNT}+k_f}{2k_f}}{1-\phi+2\phi\frac{k_f}{k_{CNT}-k_f}\ln\frac{k_{CNT}+k_f}{2k_f}},$$

$$(\rho\beta)_{nf}=(1-\phi)(\rho\beta)_f+\phi(\rho\beta)_s,\ r=\frac{\sigma_s}{\sigma_f},\ \frac{\sigma_{nf}}{\sigma_f}=1+\frac{3\left(r-1\right)\phi}{\left(r+2\right)-\left(r-1\right)\phi},$$

where $\rho_f, k_f, \mu_f, C_{pf}$, and β_f is the density, thermal conductivity, dynamic viscosity, the specific heat of the water, and thermal expansion coefficient, respectively, $\rho_s, \beta_s, k_s, \mu_s$, and C_{ps} are the respective properties of single-walled carbon nanotubes and ϕ is the solid volume fraction of carbon nanotubes.

3. Numerical Procedure

The governing momentum and energy equations (2-4) with specified boundary conditions have been solved numerically, employing the Galerkin method. To determine the solution via penalty function [39-40], we use the penalty parameter "γ" to remove the pressure P as:

$$P=-\gamma\left(\frac{\partial U}{\partial X}+\frac{\partial V}{\partial Y}\right), \tag{11}$$

The penalty parameter γ automatically satisfies the continuity equation. It is noted that for a significant value of the penalty parameter γ, it is satisfied automatically. Using Eq. (4), the momentum equation becomes:

$$U\frac{\partial U}{\partial X}+V\frac{\partial U}{\partial Y}=\gamma\frac{\rho_f}{\rho_{nf}}\frac{\partial}{\partial X}\left(\frac{\partial U}{\partial X}+\frac{\partial V}{\partial Y}\right)+\frac{\upsilon_{nf}}{\upsilon_f}\frac{1}{\mathrm{Re}}\left(\frac{\partial^2 U}{\partial X^2}+\frac{\partial^2 U}{\partial Y^2}\right), \tag{12}$$

$$U\frac{\partial V}{\partial X}+V\frac{\partial V}{\partial Y}=\gamma\frac{\rho_f}{\rho_{nf}}\frac{\partial}{\partial Y}\left(\frac{\partial U}{\partial X}+\frac{\partial V}{\partial Y}\right)+\frac{1}{\mathrm{Re}}\frac{\upsilon_{nf}}{\upsilon_f}\left(\frac{\partial^2 V}{\partial X^2}+\frac{\partial^2 V}{\partial Y^2}\right) -\frac{\sigma_{nf}}{\sigma_f}\frac{Ha^2}{\mathrm{Re}}V+\frac{(\rho\beta)_{nf}}{(\rho\beta)_f}Ri\theta. \tag{13}$$

Equations (12-13) and the related boundary conditions are solved using the Galerkin method. Consider the following solutions for velocities (U,V) and the dimensionless temperature θ using the basis set $\{\Upsilon_n\}_{n=1}^{N}$ as:

$$U\cong\sum_{n=1}^{N}U_n\Upsilon_n(X,Y), V\cong\sum_{n=1}^{N}V_n\Upsilon_n(X,Y), \theta\cong\sum_{n=1}^{N}\theta_n\Upsilon_n(X,Y). \tag{14}$$

After incorporating these solutions into Eqns. (14), we obtain the following nonlinear residual equations at all nodes of the selected cavity:

$$\mathbf{R}_k^U=\sum_{n=1}^{N}U_n\int_\Omega\left[\left(\sum_{n=1}^{N}U_n\Upsilon_n\right)\frac{\partial\Upsilon_n}{\partial X}+\left(\sum_{n=1}^{N}V_n\Upsilon_n\right)\frac{\partial\Upsilon_n}{\partial Y}\right]\Upsilon_k\mathrm{d}X\mathrm{d}Y+\gamma\sum_{n=1}^{N}U_n\int_\Omega\frac{\partial\Upsilon_k}{\partial X}\frac{\partial\Upsilon_n}{\partial X}\mathrm{d}X\mathrm{d}Y +\gamma\sum_{n=1}^{N}V_n\int_\Omega\frac{\partial\Upsilon_k}{\partial X}\frac{\partial\Upsilon_n}{\partial Y}\mathrm{d}X\mathrm{d}Y+\frac{1}{\mathrm{Re}}\sum_{n=1}^{N}U_n\int_\Omega\left[\frac{\partial\Upsilon_k}{\partial X}\frac{\partial\Upsilon_n}{\partial X}+\frac{\partial\Upsilon_k}{\partial Y}\frac{\partial\Upsilon_n}{\partial Y}\right]\mathrm{d}X\mathrm{d}Y, \tag{15}$$

$$\mathbf{R}_k^V=\sum_{n=1}^{N}U_n\int_\Omega\left[\left(\sum_{n=1}^{N}U_n\Upsilon_n\right)\frac{\partial\Upsilon_n}{\partial X}+\left(\sum_{n=1}^{N}V_n\Upsilon_n\right)\frac{\partial\Upsilon_n}{\partial Y}\right]\Upsilon_k\mathrm{d}X\mathrm{d}Y$$

$$+\gamma\left[\sum_{n=1}^{N}U_n\int_\Omega\frac{\partial\Upsilon_k}{\partial X}\frac{\partial\Upsilon_n}{\partial X}\mathrm{d}X\mathrm{d}Y+\sum_{n=1}^{N}V_n\int_\Omega\frac{\partial\Upsilon_k}{\partial X}\frac{\partial\Upsilon_n}{\partial Y}\mathrm{d}X\mathrm{d}Y\right]+\frac{\sigma_{nf}}{\sigma_f}\frac{Ha^2}{\mathrm{Re}}\sum_{n=1}^{N}V_n\int_\Omega\Upsilon_n\Upsilon_k\mathrm{d}X\mathrm{d}Y \tag{16}$$

$$+\frac{1}{\mathrm{Re}}\sum_{n=1}^{N}U_n\int_\Omega\left[\frac{\partial\Upsilon_k}{\partial X}\frac{\partial\Upsilon_n}{\partial X}+\frac{\partial\Upsilon_k}{\partial Y}\frac{\partial\Upsilon_n}{\partial Y}\right]\mathrm{d}X\mathrm{d}Y+\frac{(\rho\beta)_{nf}}{(\rho\beta)_f}Ri\sum_{n=1}^{N}\theta_n\int_\Omega\Upsilon_n\Upsilon_k\mathrm{d}X\mathrm{d}Y,$$

$$\mathbf{R}_k^\theta=\sum_{n=1}^{N}\theta_n\int_\Omega\left[\left(\sum_{n=1}^{N}U_n\Upsilon_n\right)\frac{\partial\Upsilon_n}{\partial X}+\left(\sum_{n=1}^{N}V_n\Upsilon_n\right)\frac{\partial\Upsilon_n}{\partial Y}\right]\Upsilon_k\mathrm{d}X\mathrm{d}Y +\frac{1}{\mathrm{Pr\,Re}}\frac{\alpha_{nf}}{\alpha_f}\sum_{n=1}^{N}\theta_n\int_\Omega\left[\frac{\partial\Upsilon_k}{\partial X}\frac{\partial\Upsilon_n}{\partial X}+\frac{\partial\Upsilon_k}{\partial Y}\frac{\partial\Upsilon_n}{\partial Y}\right]\mathrm{d}X\mathrm{d}Y, \tag{17}$$

The solution set of nonlinear algebraic equations (15-17) is obtained by a reduced integration method [41] and the Newton Raphson approach [42]. We get $3N \times 3N$ a system of equations at each iteration.

$$\mathbf{J}\left(\mathbf{a}^{n}\right)\left[\mathbf{a}^{n}-\mathbf{a}^{n+1}\right]=\mathbf{R}\left(\mathbf{a}^{n}\right), \tag{18}$$

where $\mathbf{J}\left(\mathbf{a}^{n}\right)$ is the Jacobian matrix and $\mathbf{R}\left(\mathbf{a}^{n}\right)$ is the residual vector.

The stream function ψ is defined as:

$$\frac{\partial \psi}{\partial Y}=U; \frac{\partial \psi}{\partial X}=-V.$$

which takes the following form

$$\frac{\partial^2 \psi}{\partial X^2}+\frac{\partial^2 \psi}{\partial Y^2}=\frac{\partial U}{\partial Y}-\frac{\partial V}{\partial X}.$$

which gives

$$\mathbf{R}_k^{\psi}=\sum_{n=1}^{N} U_n \int_{\Omega} \Upsilon_k \frac{\partial \Upsilon_n}{\partial Y} \mathrm{d}X\mathrm{d}Y-\sum_{n=1}^{N} V_n \int_{\Omega} \Upsilon_k \frac{\partial \Upsilon_n}{\partial X} \mathrm{d}X\mathrm{d}Y+\sum_{n=1}^{N} \psi_n \int_{\Omega}\left[\frac{\partial \Upsilon_k}{\partial X} \frac{\partial \Upsilon_n}{\partial X}+\frac{\partial \Upsilon_k}{\partial Y} \frac{\partial \Upsilon_n}{\partial Y}\right] \mathrm{d}X\mathrm{d}Y, \tag{19}$$

The local Nusselt number along the heating element is given by:

$$Nu=-\sum_{k=1}^{9} \theta_k \frac{\partial \Upsilon_k}{\partial Y}.$$

4. Results and Discussion

The mixed convection of nanofluid is investigated numerically in a triangular cavity including an adiabatic pentagonal obstacle. Different thermal boundary conditions are imposed on the cavity walls. The Richardson number is increased from 0.01 to 10 for different domains of convection. The solid volume fraction of carbon nanotubes is varied from 0 to 0.2, whereas the Hartmann number is increased from 0 to 50. The thermophysical properties of water and single-walled carbon nanotubes are reported in Table. 1.

Table 1. Thermophysical properties of water and single-walled carbon nanotubes [38].

Property	Water	SWCNT
ρ	997.1	2600
c_p	4179	425
k	6.2	6600
β	2.1x10^{-4}	3.3x10^{-5}
σ	0.05	3x10^{4}

The impacts of different domains of convection on streamlines, isotherms, horizontal, and vertical components of velocity are presented in Figure 2(a)-(d) for the fixed values of *Ha* and ϕ. Depending upon the values of *Ri*, Figure 2(a)-(d) present the results for the three cases. As the top wall moves with constant velocity parallel to the x-axis, the flow is generated from the right inclined wall to the left inclined wall. This flow behavior can be observed in Figure 2(a) (i)-(iii) for forced, mixed, and free convection. Due to different thermal boundary conditions, two cells form near the top surface of the triangle. The intensity of these cells increases and the size decreases with increasing Richardson number, as shown in Figure 2(a) (ii)-(iii). Finally, these cells weaken and spread at the top of the obstacle. At the same moment, another cell of high strength appears in the right bottom part of the triangle. This cell moves towards the left part with an increase in the Richardson number. The isotherms, for the same values, are presented in Figure 2(b) (i)-(iii) to show the thermal behavior of nanofluid in the cavity. Due to the higher temperature at the top surface, the heat is transferred in a downward direction. The buoyancy force decreases, and the heat transfer increases from forced to free convection, as evident from Figure 2(b) (i)-(iii). The contours of the horizontal and vertical velocity components are depicted in Figure 2(c) and 2(d) for each case. Due to the motion of the top wall, the horizontal velocity is found to be higher in the upper region of the cavity. Consequently, it pushes the fluid from the left side to the right side. This behavior can be observed in Figure 2(c) (i) for forced convection and in Figure 2 (c) (ii) for mixed convection. Due to lower velocities, in the case of free, the horizontal velocity contours are found to be weaker. These contours get stronger with an increase in the Richardson number. As the Richardson number increase, the velocity contours become stronger and move from the right side to the left side, as shown in Figure 2(c)(i)-(iii). Consequently, the upper vortices divide into more small parts. Similar behavior can be observed for the vertical velocity contour in each case. The strength of these contours increases with increasing *Ri*, as shown in Figure 2 (d) (i)-(iii).

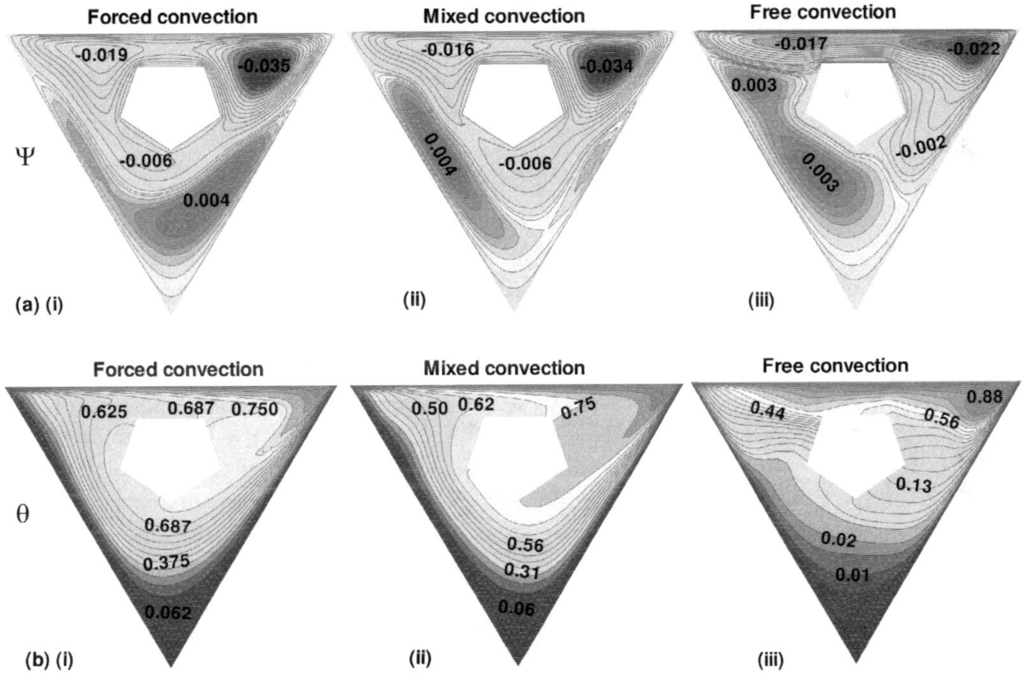

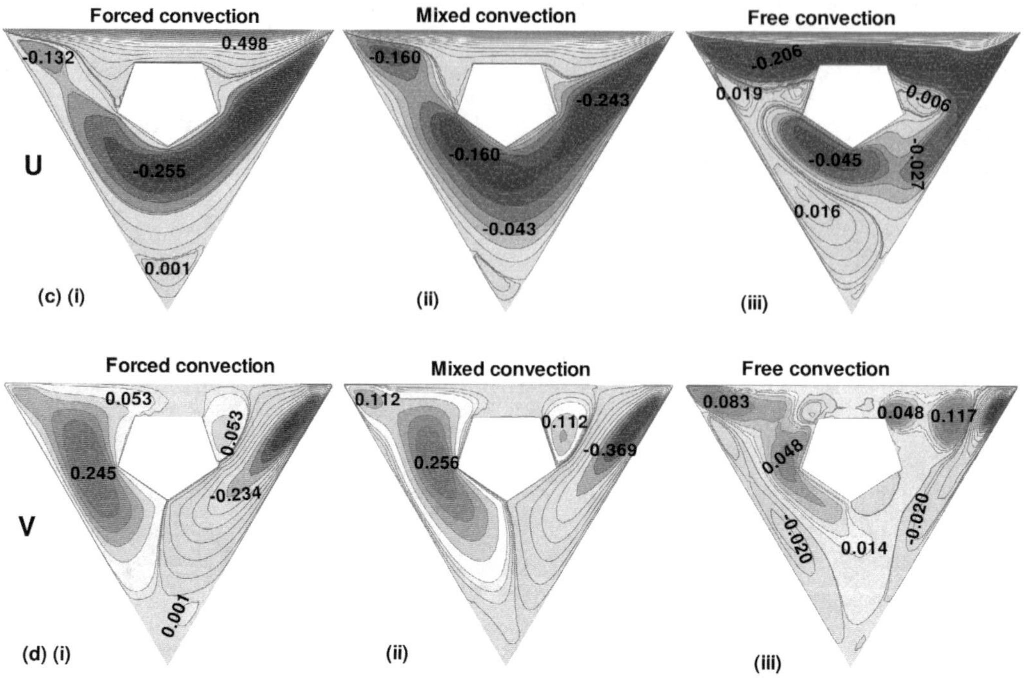

Figure 2. Variation of streamlines and isotherms for different Richardson numbers with Re = 500, Ha = 10 and ϕ = 0.1.

The 2-D plots for the dimensionless velocity components, temperature, and heat transfer rate are presented in Figure 3(a)-(e) for ϕ=0.1 and Ha=10. The behavior of the horizontal and vertical velocities for the three cases is shown in Figure 3(a) and (b) along the horizontal/vertical lines. As expected, the horizontal velocity is zero at the corner and on both sides of the pentagonal obstacle in case of free convection. Within the cavity, the horizontal velocity decreases/increases due to change in the boundary conditions.

Similarly, in the case of mixed convection, the horizontal velocity increases/decreases in the lower empty regions of the cavity respectively. Due to different thermal boundary conditions applied on the cavity walls and the heated obstacle, the horizontal velocity magnitude is higher on both sides of the obstacle, as shown in Figure 3(a). The vertical velocity, in case of mixed and free convection, is found to be negligible on the left side. However, in the right part, the magnitude of the vertical velocity increases due to an increase in the Grashof number, Figure 3(b). The behavior of the dimensionless temperature on both sides of the obstacle is shown in Figure 3(c) for all three cases of convection. Due to obstacle, there is a discontinuity in the temperature and then increases in the free space on the right side. It is essential to note that this behavior is different on both sides. In the left region, the dimensionless temperature increases in the case of free convection, whereas for mixed and forced convection, a negligible rise of temperature was observed. In the right region, the temperature goes to the maximum due to the movement of the top wall in the right direction. In the stable region, the temperature increases with increasing Ri. The variation in Nu_{loc} along the heated top surface and around the adiabatic obstacle is shown in Figure 3(d)-(e), respectively, for the three selected cases. The local Nusselt number decreases along the horizontal surface. At both corners of the top surface, there is a sudden increase or decrease in the local Nusselt number. The maximum values of Nusselt numbers are obtained at the corners of the heater due to singularity at those points. However, along with the adiabatic pentagonal obstacle, the variation in the local Nusselt number is strange, as shown in Figure 3(e).

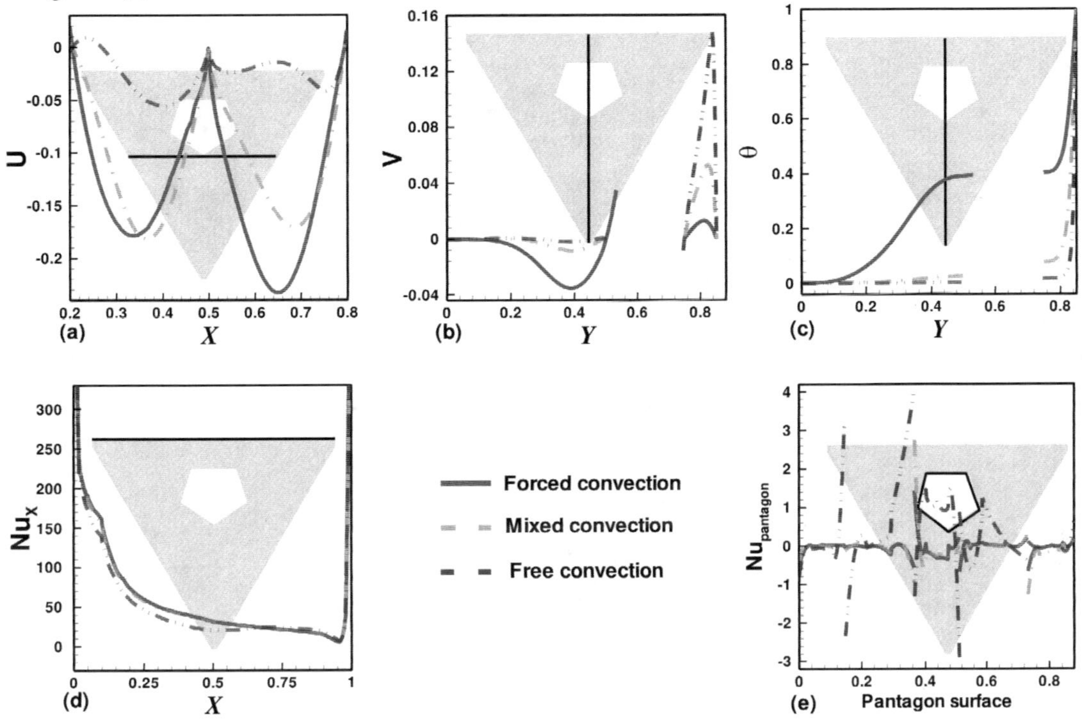

Figure 3. Effects of Richardson number on dimensionless velocities, temperature, and local Nusselt numbers with $Re = 500$, $Ha = 10$ and $\phi = 0.1$.

The variation of isotherms of nanofluid in the cavity with the volume fraction of carbon nanotubes is demonstrated in Figure 4(a)-(c) for the three types of convection. In the case of forced convection, for pure water $\phi=0$, the heat transfers slowly from the heated top surface to the right side, Figure 4(a)(i). The isotherms are equally spaced around the adiabatic obstacle. The heat transfer rate increases with an increase in the volume fraction of carbon nanotubes, Figure 4(a) (ii)-(iii). In the case of mixed and free convection, an increase in heat transfer can be observed with an increase in the volume fraction of carbon nanotubes, see Figure 4(b)-(c) (i)-(iii). The variation of dimensionless temperature and local Nusselt numbers with volume fraction is also presented in Figure 5(a)-(c) for different values of the Richardson number. The dimensionless temperature increases vertically in the case of forced and mixed convection, Figure 5(a)-(b) respectively. In the case of forced convection, the negligible effect of the solid volume fraction of nanotubes on the dimensionless temperature can be noticed in the left region of the cavity. However, in the right-side space, a considerable effect of the solid volume fraction of carbon nanotubes can be observed. The variation of the local Nusselt number along the horizontal heated surface is depicted in Figure 5(b) (i)-(iii). In each case, a sudden change in the local Nusselt number is noticed due to singularity at the corners. As expected, the local Nusselt numbers are lower for the pure water and increase with increasing solid volume fraction of nanotubes.

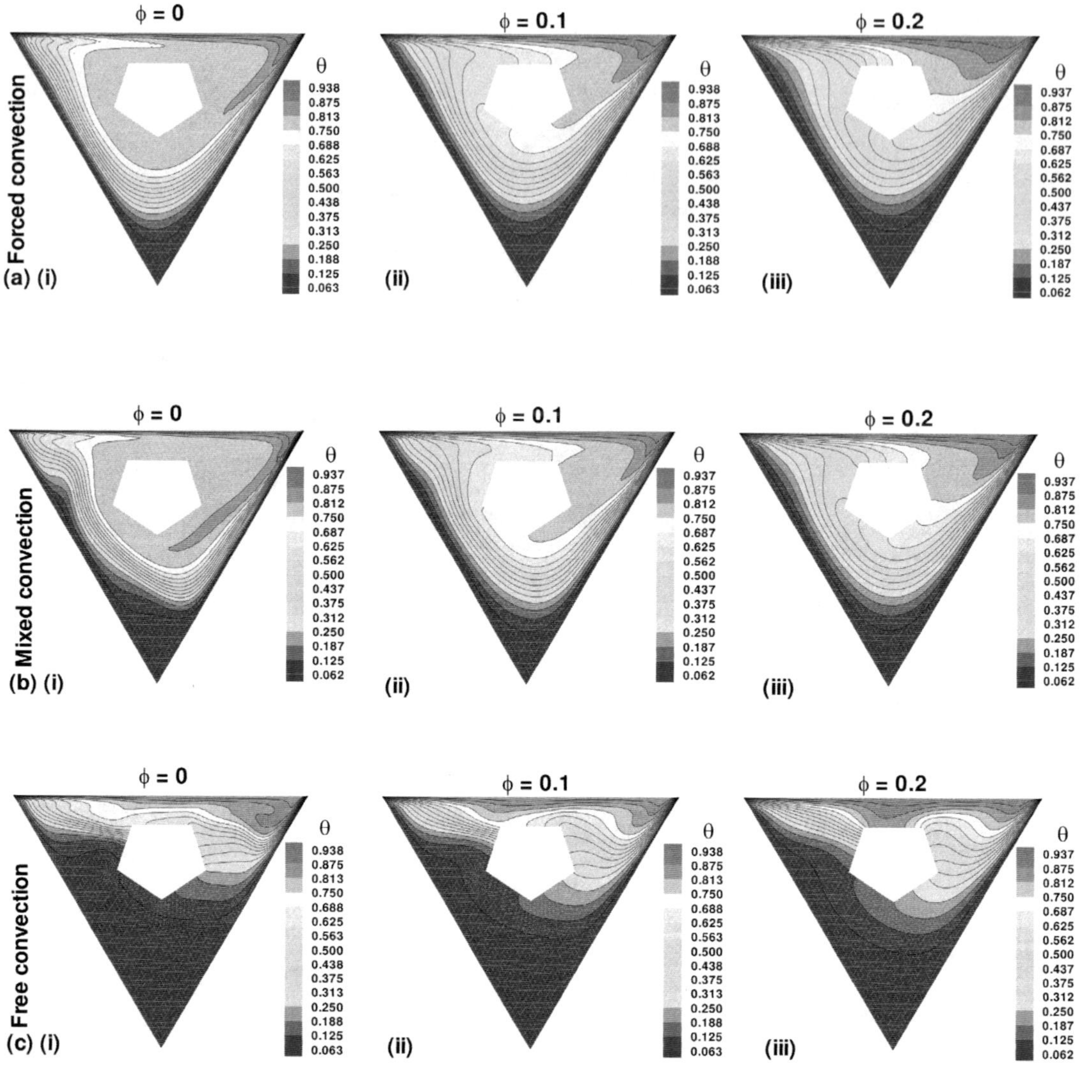

Figure 4. Effects of volume fraction of carbon nanotubes on isotherms for different Richardson numbers with $Re = 300$ and $Ha = 10$.

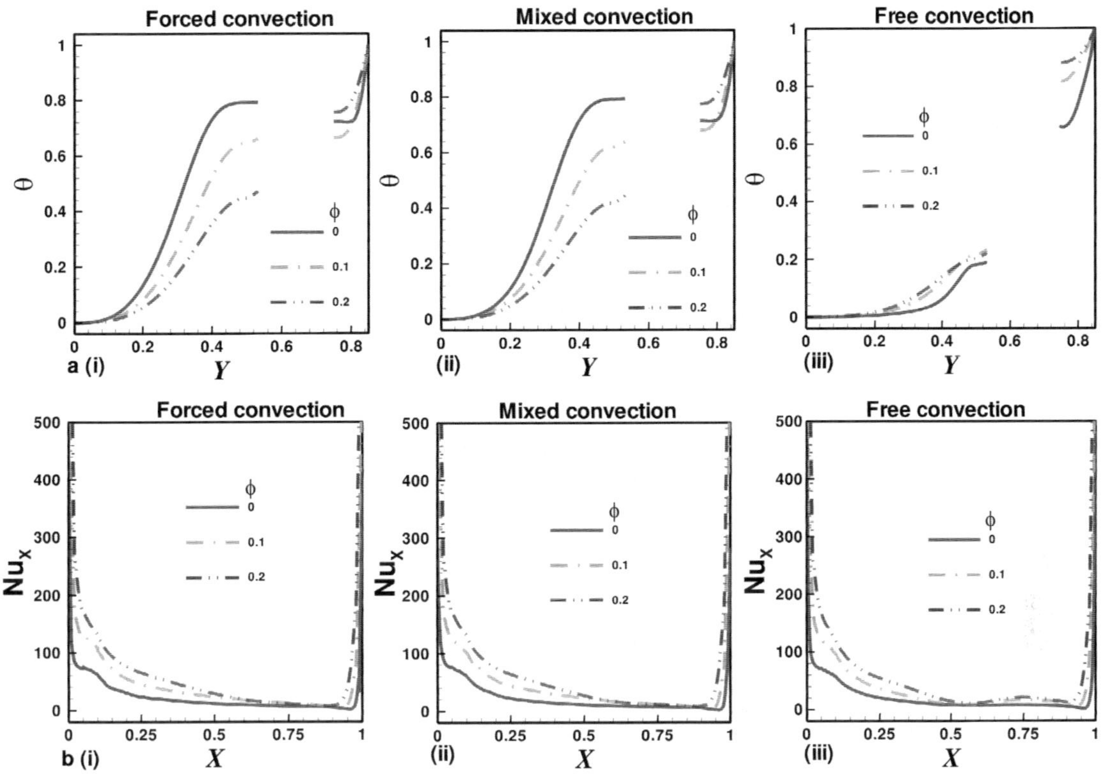

Figure 5. Effects of volume fraction of carbon nanotubes on (a) vertical temperature and (b) Nusselt number for different Richardson numbers with Ha = 10 and Re=200.

The influences of Ha on the isotherms for the three domains of convection are depicted in Figure 6(a)-(c) keeping other parameters fixed. When Ha=0, the isotherms are shown in Figure 6(a) (i) for forced convection. As the top surface moves towards the right, the heat is transported slowly from the heated surface to the colder sides and the isotherms are equally spaced. The heat transfer increases with an increase in the magnetic field. Since the magnetic field produces Lorentz forces, which resists the motion of the nanofluid, this fact is explained in Figure 6(a) (ii)-(iii). The same behavior can be observed in Figure 6(b) and (c) for mixed and free convection. As expected, the inertia forces are dominant in case of forced convection, which increases the Nusselt number.

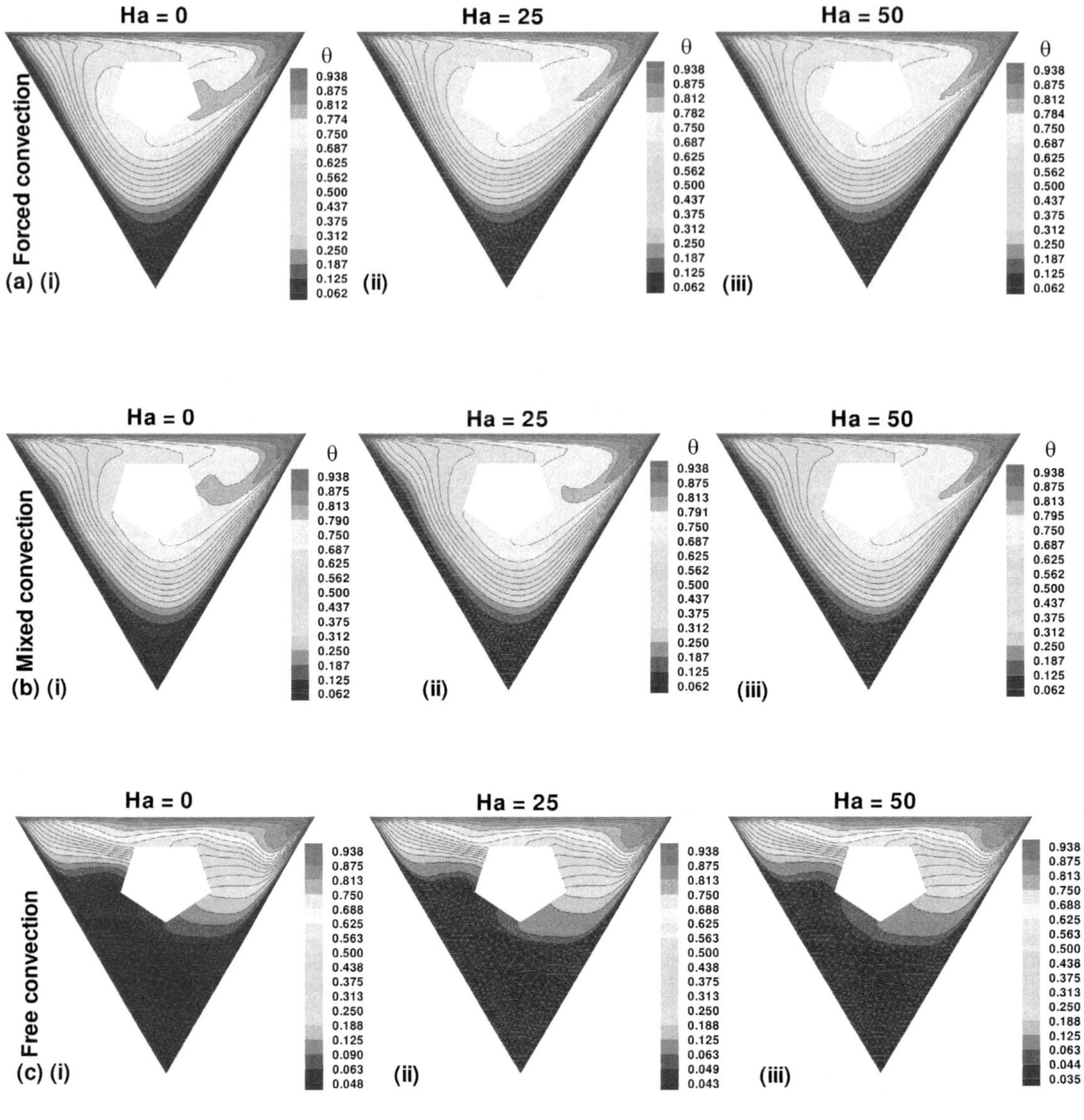

Figure 6. Effects of Hartmann number on isotherms for different Richardson numbers with ϕ= 0.1 and *Re*=500.

The effects of the volume fraction of carbon nanotubes on the average Nusselt number are presented in Figure 7(a) for the three selected domains of convection. In this case, the magnetic field is kept constant. As expected, the heat transfer rate increases with the increasing volume fraction of nanotubes. This is because carbon nanotubes show higher thermal conductivity (Table 1).

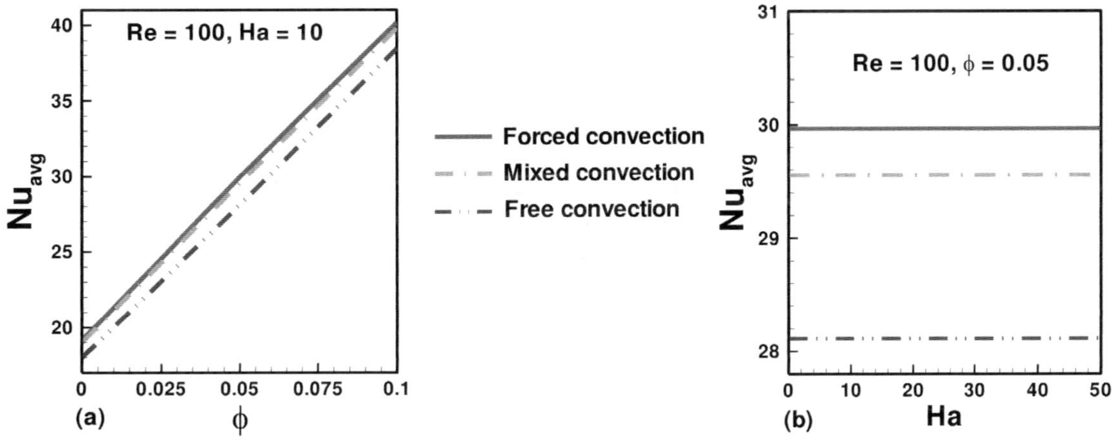

Figure 7. Effect of volume fraction of carbon nanotubes and Hartmann number on average Nusselt number for different Richardson numbers.

Consequently, it increases the thermal conductivity of the nanofluid and the heat transfer rate increases. The effects of smaller values of the Richardson number show the negligible effect on the Nusselt number. However, forced convection shows a significant impact on the average Nusselt number. The variation of average Nusselt number with Hartmann number is depicted in Figure 7(b) for three selected Richardson number. In this case, the volume fraction of carbon nanotubes is kept constant. It is observed that, for each type of convection, the average Nusselt number practically remains constant. For free convection, the average Nusselt number is found to be the lowest.

5. Conclusion

The steady mixed convection of carbon nanotubes mixed with water is investigated numerically in a triangular cavity in the presence of an adiabatic pentagonal obstacle. The effects of the volume fraction of carbon nanotubes and the magnetic field are examined for the three domains of convection. It is concluded that

- The streamlines and isotherms explain the flow and heat transfer in the cavity.
- The heat transfer enhances due to the pentagonal obstacle in the cavity.
- The heat transfer rate is highest in case of forced convection.
- The Hartmann number reduces the heat transfer rate, whereas the volume fraction of nanotubes helps in improving the heat transfer rate in the cavity.
- The Richardson number reduces the heat transfer rate in the lid-driven cavity with a pentagonal obstacle.

6. References

[1] Selimefendigil F 2019 *J Propul Power* **8** 128-137.
[2] Jafari M, Farhadi M and Sedighi K 2014 *Trans. Phenom. Nano Micro Scales* **2** 14-28.
[3] Öztop H F, Sakhrieh A, Abu-Nada E, Al-Salem K 2017 *Int. Commun Heat Mass* **86** 42-51.
[4] Selimefendigil F and Öztop H F 2016 *Int. J. Mech. Sci.* **118**, 113-124.
[5] Selimefendigil F and Öztop H F 2017 *J. Taiwan Inst. Chem. E.*, **70** 168-178.
[6] Sivasankaran S, Sivakumar V and Prakash P 2010 *Int. J. Heat Mass Tran.* **53** 4304-4315.
[7] Mekroussi S, Sahraoui K, Mebarki B and Benchattir A 2017 *Int. J. Heat Tech.* **35** 1023-1033.
[8] Sivasankarana S, Ananthanb S S and Hakeem A K 2016 *Sci. Iran.*B, **23** 1027-1036.
[9] Sivakumar V and Sivasankaran S 2014 *J. Appl. Mech. Tech. Phy.* **55** 634.
[10] Sivasankaran S and Pan K L 2012 *Numer. Heat Tr A-Appl.* **61** 101-121.

[11] Bakar N A, Karimipour A and Roslan R 2017 *AIP Conference Proceedings*
[12] Manchanda M. and Gangawane K. M. 2018 *Int. J. Mech. Sci.* **144** 235-248.
[13] Gangawane K M and Manikandan B 2017 *Chin. J. Chem. Eng.* **25** 1381-1394.
[14] Gangawane K M 2017 *Int. J. Heat Mass Tran.* **105** 34-57.
[15] Kalteh M, Javaherdeh K and Azarbarzin T 2014 *Powder Technol.* **253** 780-788.
[16] Ali M M, Alim M A and Ahmed S S 2017 *Procedia Eng.* **194** 479-486.
[17] Louaraychi A, Lamsaadi M, Naïmi M, El-Harfi H, Kaddiri M, Raji A and Hasnaoui M 2019 *Int. J. Heat Mass Tran.* **132** 394-406.
[18] Izadi S, Armaghani T, Ghasemias R, Chamkha A J and Molana M 2019 *Powder Technol.* **343** 880-907.
[19] Ezzaraa K, Bahlaoui A, Arroub I, Raji A, Hasnaoui M and Naïmi M 2019 *Int. J. Mech. Sci.* **153–154** 310-320.
[20] Ghasemi B and Aminossadati S M 2010 *Int. Commun. Heat Mass* **37** 1142-1148.
[21] Nayak R K, Bhattacharyya S and Pop I 2018 *Int. J. Heat Mass Tran.* **125** 908-919.
[22] Yu Q, Xu H and Liao S 2018 *Int. J. Heat Mass Tran.* **126 Part B** 221-236.
[23] Javed T and Siddiqui M A 2018 *J. Mol. Liq.* **249** 831-842.
[24] Kapil M, Roy D, Sharma B, Rana S C, Pramanik S and Barman R N 2019 *Materials Today: Proceedings* **11** 700-707
[25] Al-Rashed Abdullah A A A, Kalidasan K, Lioua K, Velkennedy R, Abdelkarim A, Ahmed K H and Emad H M 2018 *Int. J. Mech. Sci.* **135** 362-375.
[26] Wang T, Wang Z, Xi G and Huang Z 2017 *Int. J. Heat Mass Tran.* **111** 1250-1259.
[27] Krunal M, Gangawane B, Manikandan Z 2017 *Chin. J. Chem. Eng.* **25** 1381-1394,
[28] Selimefendigil F and Öztop H F 2019 *Physica A* **534** 122-144.
[29] Mojtabi A, Khouzam A, Yacine L and Marie-Catherine C M 2019 *Int. J. Heat Mass Tran.* **139** 1037-1046.
[30] Jahirul H M M., Alim M A, Bhuiyan A H and Ali M 2017 *Procedia Eng.* **194** 442-449.
[31] Arani A A A, Ababaei A, Sheikhzadeh G A and Aghaei A 2018 *Alex. Eng. J.*, **57** 1287-1300.
[32] B. Karbasifar, M. Akbari, D. Toghraie 2018 *Int. J. Heat Mass Tran.* **116**1237-1249.
[33] Kefayati G H R and Tang H 2018 *Int. J. Heat Mass Tran.* **124** 344-367.
[34] Pal S K, Bhattacharyya S and Pop I 2018 *Int. J. Heat Mass Tran.* **127 Part A** 885-900.
[35] Salar T and Azin A 2018 *Int. J. Therm. Sci.* **134** 242-257.
[36] Kashyap D and Dass A K 2019 *Int. J. Mech. Sci.* **157** 45-59
[37] Astanina M S, Sheremet M A, Oztop H F and Abu-Hamdeh N 2018 *Int. J. Heat Mass Tran.* **118** 527-537
[38] Khan W A, Khan Z H and Rahi M 2014 *Appl Nanosci* **4-5** 633-641
[39] Basak T, Roy S, Sharma P K and Pop I 2009 *Int. J. Heat Mass Tran.* **52** 2224-2242
[40] Reddy J N 1993 *An Introduction to Finite Element Method* McGraw-Hill, New York.
[41] Zienkiewicz O C, Taylor R L and Too J M 1971 *Int. J. Numer. Methods Eng.* **3** 275-290.
[42] Roy S and Basak T 2005 *Int. J. Eng. Sci.* **43** 668–680.

AUTHOR INDEX